An Alternative to the Standard Model of Physics

An Alternative to the Standard Model of Physics

◆

The Waves of Particles Theory of Light

John E. Royer

iUniverse, Inc.
New York Lincoln Shanghai

An Alternative to the Standard Model of Physics
The Waves of Particles Theory of Light

Copyright © 2007 by John E. Royer

All rights reserved. No part of this book may be used or reproduced by any means, graphic, electronic, or mechanical, including photocopying, recording, taping or by any information storage retrieval system without the written permission of the publisher except in the case of brief quotations embodied in critical articles and reviews.

iUniverse books may be ordered through booksellers or by contacting:

iUniverse
2021 Pine Lake Road, Suite 100
Lincoln, NE 68512
www.iuniverse.com
1-800-Authors (1-800-288-4677)

Because of the dynamic nature of the Internet, any Web addresses or links contained in this book may have changed since publication and may no longer be valid.

The views expressed in this work are solely those of the author and do not necessarily reflect the views of the publisher, and the publisher hereby disclaims any responsibility for them.

ISBN: 978-0-595-44092-4 (pbk)
ISBN: 978-0-595-88416-2 (ebk)

Printed in the United States of America

Contents

Acknowledgments vii
Preface ix

Chapter 1   A New Theory of Electromagnetic Radiation 1
    Light as Waves of Particles 2
    A Different Way of Viewing EMR 4
    The Four Types of Electromagnetic Radiation 7
    Neutral Electromagnetic Radiation 8
    How EMR Is Radiated and What It Is like during Transmission 11
    The Alternating Spins in Waves of Basic Particles 12
    Solar Sailing 14
    Refraction in the Waves of Particles Theory of Light 17
    Another Effect of the Refraction of Light 20
    The Waves of Particles Theory and the Universe 21

Chapter 2   The Origin of Matter and the Electromagnetic
                  Force 25
    In the Beginning 26
    A Universe Blossoms 27
    Matter: Electrons and Protons 28
    The Origin of the Electrical Charge 29
    The Origin of the Magnetic Force 33
    Opposite Magnetic Poles 34
    The Neutron 36

Chapter 3   The Forces of Nature 39
    The Weak Nuclear Force 40
    The Strong Nuclear Force 43

Quantum Tunneling of Alpha Particles	47
The Atom	47
Molecules	51
Gravity	51
CHAPTER 4 Conclusions	57
General Relativity and Quantum Mechanics	58
References	63

Acknowledgments

This new theory would have never been possible without the computer and the Internet. The wealth of information available because of these two inventions is overwhelming.

Special thanks to my brother Paul, who provided an incredible amount of articles from the Internet, and to my sister Esther, who provided her expertise and advice on the English language.

Preface

Originally, this book was going to be about a new theory of electromagnetic radiation. It wasn't long before it became apparent that this new view on light would also be able to describe and explain many of the mysteries the Standard Model could not, such as the origin of the electrical charge, magnetism, time, and mass. This book also gives a comprehensive explanation as to how gravity works and proposes a test to verify or refute the theory of gravity.

Parts of this book may be repetitive and boring to some readers. Feel free to skip those parts if you like, since each chapter explains most of the basics anyway. Then, when you are finished, go back and read the parts you skipped or reread sections that were hard to follow.

◆ ◆ ◆

Footnotes and sources are kept to a minimum, mainly because sources found on the Internet are frequently here today and gone tomorrow. So when you read this theory, you may encounter statements of fact that you may want to verify. If you have questions, check out several different sources.

◆ ◆ ◆

There is no question as to whether this theory has errors in it. It is a work in progress—as any scientific work is—and improvements will only come with challenges and refinements in the future.

1
A New Theory of Electromagnetic Radiation

Light as Waves of Particles

Could electromagnetic radiation (EMR) be waves of particles instead of waves of light or photons? Waves of light, in the classical sense, need something to propagate through—such as ether. Waves of a basic particle, like virtual particles transmitting EMR, would not need ether or anything else to travel through. EMR as waves of particles could also eliminate all the problems associated with the photon theory of light. There are better ways to explain the particle-like (photon) characteristic of light. The photon theory of EMR is one reason the Standard Model and our present view of the universe is in such a complicated, contradictory state. Until we have the correct theoretical understanding of EMR, we cannot hope to understand the universe in front of our eyes.

Consider the following: a photon is Planck's constant times the frequency of the radiation. What about the momentum and energy a photon of light contains? Momentum is mass times velocity, and energy is 1/2 times mass times velocity^2. A photon has no mass, and its velocity is fixed by natural law, so the momentum and energy a photon contains is directly related to its frequency. Frequency is directly related to the length of a man-made unit of time, the second. If the length of a second is changed, the momentum and energy a photon contains changes also. This is because there are now more or fewer waves in each second, and this changes the momentum and energy of the photon.

All of our arbitrary units of measurement must be measured and conform to the basic units of nature. So how can a photon be a naturally occurring particle? The momentum and energy of a photon conforms to the number of waves in a man-made second. The odds are astronomically against the momentum and energy of a photon of light being exactly equal to the length of a man-made unit of time. If this is true, then a photon cannot be a naturally occurring particle and can only be useful as an energy and momentum measurement of EMR, similar to a foot-pound or watt and other measurements.

Now, how can one explain the particle-like characteristic of EMR, especially in the x-ray and gamma ray range of the electromagnetic spectrum? Even at the extremely high frequencies in the gamma ray range, EMR is a wave of particles, and not a naturally occurring particle called the photon. True, it certainly seems to behave like a particle, but that does not make it a particle. The particle-like characteristic comes from its high frequency and the pulse-like way electrons emit radiation. [57] One fact no one can ignore is that the higher the frequency of the EMR, the more particle-like it becomes. At radio wave frequencies, EMR behaves almost entirely like a wave, and at gamma ray frequencies, entirely like a particle.

One way to view this is to imagine someone slapping you ten times in a second. You would describe it as a volley or series of slaps, or a wave of slaps. What if this person (if it was physically possible) slapped you one hundred times in a second? You would no longer describe it as a wave of slaps. Since our senses cannot feel anything in much less than a twentieth of a second, you would no longer feel the individual slaps. Instead you would say you were slugged, hit once, like a particle.

This is similar to what happens in the photoelectric effect. [1] The electron is ejected when a frequency high enough to break the electron's bond to the atom is absorbed in a short enough time period. Lower frequencies cannot break this bond, because the electron simply passes the energy to the rest of the atom or radiates it away before it ever achieves a high enough energy level to break its electromagnetic bond with the nucleus. This is true with all high-frequency radiation. EMR never becomes a particle (hf) until gamma rays at a frequency high enough strike a nucleus and create an electron and a positron, but then it is no longer EMR. [4]

Recent photoelectric experiments with laser light have shown that lower frequencies of EMR can in fact eject electrons from the gold foil experiment. [1, 2, and 3]

It is tempting and convenient to consider EMR as a particle, because the energy levels from which the hydrogen atom radiates EMR differ by the energy quanta in a photon, but that is the only atom for which the photon quantity of emitted radiation holds true. After saying all this, the photon theory of light does have merit. Light, or EMR, is radiated and, to a lesser extent, absorbed in particle-like quanta close to but not exactly photon quantities of hf. It is radiated in short pulses of superwaves training each other in the form of wavetrains. [3] Each superwave contains about a billion light waves and each wavetrain or pulse of light is about a foot long. (Superwaves will be explained shortly in the next section.) The length of a pulse of radiation varies significantly with longer wavelengths having longer pulses and shorter wavelengths having shorter pulses or wavetrains. Higher frequencies of radiation, in addition to having a shorter wavelength, also have a narrower wave front. This is the major reason higher frequencies of EMR are more penetrating than lower frequencies.

Adding all these features of EMR together, we have higher frequencies and shorter pulses with a narrower wave front that accounts for the particle-like characteristic of EMR as the frequency is increased. Note that it is particle-like, not a particle. High frequencies of EMR, especially in the gamma ray range, are almost exclusively particle-like, but they are not particles in the classic sense like electrons or protons. In this theory, the use of the word photon means particle-like quantities of EMR (hf). A pulse of EMR ejected when an electron or proton falls to a lower energy level may also be referred to as a photon in this theory. [56]

There are additional problems with the photon concept of EMR. It is the only particle commonly known that has no defined length or width. It can be a thousand miles long and at least ten meters wide—or as wide as the largest telescopes in use today. A photon is just a collection of EMR equal to hf in the waves of particles theory of light.

Now, let's return back to the classical wave theory of light. Here there are also problems describing some of the characteristics of light. The first is how light striking a surface can be transmitted, absorbed, and reflected all from the same beam of light. This is also a problem for the photon theory of light. Another problem is how light passes through a material such as a thin sheet of silver when the wavelength is longer than any holes in the material. [5, 6 and 7] The photon theory cannot explain this either. Are the photons ripped apart and then reassembled to create whole physically larger photons that emerge from the other side? The Wave of Particles Theory of Light (WPTL) can answer these questions and more. It can also explain why all particles with mass have wave- and particle-like characteristics.

An interesting thing about the photon theory of light that is never touched on is how a photon changes frequency, upon reflection, like in the Compton Effect. [13] We know EMR is emitted in waves, so what happens to the missing waves in the Compton Effect? How does nature decide how many waves to eliminate? Later, we will see that none of the waves are eliminated in the Compton Effect; this creates a major problem for the photon theory of light.

So is EMR a wave or a particle? Neither! It is waves of particles much more fundamental than the photon.

A Different Way of Viewing EMR

The Wave of Particles Theory of Light (WPTL) may be a little difficult to understand. A person may ask why bother with it? The complexity of this theory will be returned a thousand-fold in the beauty and simplicity of understanding the universe.

There is one way to describe EMR that can answer most, if not all of the seemingly bazaar characteristic light displays: waves of particles with the following characteristics:

(1) Each particle of light contains 1/2 Planck's constant of energy (1/2h). The reason it is half is because each wave of light contains two particles per frequency spaced half wavelength apart. The two particles are identical except for spin, which alternates and gives EMR its alternating electromagnetic fields. The size of these particles is the smallest possible in nature and cannot have a cross-section of more than a quarter the wavelength of the highest frequency of any possible gamma ray.

(2) Electrons emitting radiation eject millions to billions of these particles per oscillation out away from the nucleus. The actual number is assumed to be directly related to the type of atom radiating the EMR, as well as the frequency being radiated. The number is assumed to be the direct inverse to the frequency being radiated per oscillation. This may seem counterintuitive, but remember lower frequencies of EMR are radiated by electrons containing more kinetic energy than electrons radiating higher frequencies, such as x-rays. Also, the lower the frequency, a wider wave front carries away more particles, and therefore more energy. A graphical way of looking at this is to assume this series of periods ... are x-rays superwave fronts and this series of parentheses) ------) -------) ------) (hyphens for spacing only) are infrared light superwaves. It is easy to see that the wave front of the x-rays is much smaller, and therefore also much more penetrating than the infrared superwave front. The width of a superwave front is close to the wavelength.

(3) The electromagnetic field that surrounds electrons is what is being radiated away at the velocity of light. Magnetic fields are composed of these basic particles with the same spin. Electric fields are basic particles with their angular momentum lined up as to have the angular momentum all facing the same direction at the same moment in time. You could view the basic particles as rotating strings with the loop part of the strings all rotating like dancers in unison. In fact, this is the origin of the electric force and spin is the origin of the magnetic force. This view can explain why the two are intricately tied together; they originate from the same particle, but at the same time, you can have one of the fields without the other.

(4) These particles could be viewed as virtual particles that pop into existence, but do not go out of existence until the radiation is absorbed. This is because the oscillating electron accelerates them to the velocity of light. You could say time ceases for the virtual particles because at the velocity of light time stops. This theory states the particles are real and exist in nature in astronomically larger numbers than any other particle.

(5) The four types of EMR are:

    (a) Single electron emissions—light, x-rays
    (b) Multiple electron emissions—radio waves
    (c) Proton emissions from the nucleus—gamma rays
    (d) Neutron emissions from the nucleus—neutral EMR, neutrinos, neutral radiation (More later on this neutral radiation.)

A way of looking at EMR and to keep it simple for transmission over the Internet would be to view each:

(1) Individual particle as either < or >, which represents the possible spin states of each half-wave. Each > is one particle equal to 1/2h.

(2) Mass ejections from an oscillating electron as))). Each parenthesis represents a half superwave of approximately one billion basic particles.

A wave from here on will refer to a Planck's constant worth of energy—as in the classical wave theory of light—and a superwave as an electron oscillating twice ejecting approximately one billion waves at one h per wave [< >]. The number of waves [< >] per superwave [))] is a very imprecise number. As a general rule of thumb, think of more at lower frequencies and less at the higher x-ray range. The same frequency of EMR may contain a wide range of waves [< >] per superwave [))] depending on the type of element radiating, as well as other factors such as possibly even pressure on the radiating atoms.

To visualize how miniscule these basic particles are we refer to Planck scales. There are about 10^{43} Planck quanta seconds in one second. This equals the number of Planck lengths in the distance light travels in one second. So there must be about 10^{43} Planck lengths in 300,000 kilometers. [29]

What we have is a wave (1h) of EMR, which is two basic particles containing 1/2h of energy or momentum in sequence, like this < >, and a superwave of two billion virtual particles radiated out of an atom in two succeeding half wave fronts, like this)). Basic particles in half a superwave have the same spins when radiated and absorbed by atoms, but often do not have the same spins when traveling from atom to atom. Later, we will explain why superwaves of these basic or virtual particles always radiate from the electron and are absorbed with alternating spins while the basic particles surrounding the electron all have the same spin. A half superwave [)] is a collection of approximately one billion 1/2h particles [>] magnetically connected to each other. It can be viewed as a piece of magnetic flux radiated from an electron because that is exactly what it is in this theory.

An interesting experiment—that cannot be explained very well by the classical wave or photon theory of light—is the thin silver sheet experiment mentioned earlier. [5] Some of the light goes through this thin silver sheet. The light has a wavelength longer than any of the holes for it to go through. Therefore it should be either reflected or absorbed. In the WPTL, the superwaves are broken into shorter segments. Some are absorbed, some reflected, and some of the shortest segments are transmitted through to the other side. A single electromagnetic half-wave or light particle has a cross-section of a Planck length squared.

The Four Types of Electromagnetic Radiation

There are four types of electromagnetic radiation: single electron emission, radio waves, gamma rays and neutral radiation. **Single electron radiation** was just explained.

Radio waves are nothing more than multiple single electron emissions. This can be as few as two in the infrared region—to no known upper limit. Single and multiple electron emissions probably overlap in the infrared to the microwave region of frequencies. Assume a radio broadcasts with the emissions of 10^{28} plus of single electron emissions in each cycle of superwaves. Imagine a radio wave with a wavelength of one meter being broadcasted. The wave is radiated from an antenna also one meter long (this is not necessary in the WPTL). As the electric current is forced to oscillate up and down the antenna, a cycle of superwaves (per frequency) is radiated away in all directions. Each of these cycles, shaped much like the radio waves we have seen drawn many times before, contains 10^{28} of single electron superwave emissions. Each superwave in turn contains about a billion waves of h, Planck's constant worth of energy.

So we have each frequency (cycle), containing 10^{28} of superwaves, with each containing about a billion waves of EMR. Radio waves get into astronomical numbers of waves (< >) per cycle or frequency. The number is directly related to the power one wants or is permitted to broadcast with no theoretical upper limit.

Radio waves are multiple single electron emissions of superwaves [ ))))] magnetically connected in the shape of the classical view of undulating waves. Radio superwaves are broken into smaller magnetically connected segments continually in transit by trees, buildings, and even antennae. Even atoms and dust particles in the atmosphere continually break radio waves into smaller segments.

This view of radio waves explains why a seemingly unlimited number of radios can receive a single broadcasted radio signal and why a radio wave can be reflected, absorbed (received), and passed through a given area all at the same time. This makes it much easier to understand how EMR can be absorbed, reflected, and/or transmitted by the same material under the same conditions. Parts of superwaves can travel through, between, and reflect off atoms. A dish-shaped antenna simply reflects these individual waves into a focal point, concentrating them into a receivable amount, or into photons. A telescope does the same thing to widely dispersed light waves from distant parts of the universe. It concentrates the light waves into a focal point of usable quantities, such as a photon of light.

Gamma rays are radiated from the protons contained in the nucleus of an atom. (High velocity electrons can also radiate gamma rays.) Both electron emission and proton emission of EMR radiate away the exact same particles—virtual or basic particles—that contain exactly the same amount of energy: 1/2h per particle. Each wave contains exactly the same amount of energy: 1h per two particles. Waves of gamma rays are exactly like every other frequency of EMR. The difference is in the frequency and the number of waves (h) or particles in each cycle (frequency). Whereas earlier we mentioned that each superwave of electron emission contained about one billion waves (h) or two billion particles in the light wave frequency range and perhaps 1/1000 that amount in the x-ray range of frequencies, gamma ray proton emissions radiate away approximately one thousand times as much energy per superwave when compared to x-rays of the same frequency. It is assumed the proton radiates away more per oscillation at the same frequency when compared to an electron. This is because of the proton's larger size and mass. This is another reason gamma rays behave almost exclusively as a particle; the wave characteristic seems to disappear. As mentioned earlier, the higher the frequency being radiated, the shorter the pulse or wavetrain of EMR radiated. A single gamma ray wavetrain probably has a length in the millimeter or shorter range. This adds further to the particle-like nature of higher frequencies like gamma rays.

Neutral Electromagnetic Radiation

Now onto neutral EMR. This sounds like a contradiction in terms—and in a way it is—but neutrinos are neutral radiation. They contain exactly the same amount of energy per wave, and probably slightly more per superwave compared to gamma ray radiation. Neutrinos should also be considered radiation because they are radiated and absorbed in the same manner—that is, they can only be absorbed by a particle that is capable of radiating it. This means only neutrons can absorb energy transmitted by neutrino radiation. This is one reason why neutrinos so rarely interact with matter. Neutrino radiation may also be reflected by neutrons, protons, and electrons. There are differences:

1. They are radiated away by neutrons or other particles with mass that have positive and negative electric fields around them.
2. The electromagnetic field of each wave does not alternate, because the two alternating basic particles of the half-waves are entangled into a single neutral unit.

In regular EMR the waves are like this: <----->-----<----->. In neutrino radiation the two particles become one neutral particle like this: <>----------<>. (Hyphens for spacing only.) Neutrons actually radiate away the neutral fields that surround them when they are placed in a higher, but unstable, energy level in the nucleus, just as electrons and protons radiate away the charged fields around them. There are three ways this can come about: (1) the nucleus is struck by a particle; (2) gamma rays strike and are absorbed by a proton in the nucleus, which transfers energy to a neutron; and (3) the neutron absorbs neutrino radiation. Neutrons radiate this neutral neutrino radiation at extremely high frequencies in the gamma ray range.

Therefore, neutrino radiation is almost exclusively particle-like. In fact, this is the main reason it is considered a particle. Also, neutrino radiation cannot be detected because it does not create alternating electromagnetic fields when absorbed, reflected, or transmitted through electromagnetic or neutral fields. Like higher frequencies of EMR, neutrino radiation is radiated in short pulses that contribute greatly to its particle-like characteristic.

Thinking about this new view of the neutrino, one thing stands out: it can explain the major source—but not the only source—of radioactive decay of a nucleus. There is no need for a weak nuclear force or the W and Z particles. There are some major differences from the present view of neutrinos. First, neutrinos have no mass. Second, they interact with matter more than the present theory allows for, perhaps ten parts per million more. The only difference between the electron neutrino and its cousins, the muon and tau neutrinos, is in the concentration of the neutral waves; that is, the number of waves radiated or absorbed in a given amount of space-time. They are radiated in wavetrains from the nucleus. But due to the movement of the radiating and absorbing neutrons, parts of wavetrains are many times absorbed. These parts of wavetrains may not have enough energy to cause a nucleus to decay. Therefore, there may be no obvious evidence of this absorption. The only evidence may be an increase in the temperature of the nucleus, which soon transfers this temperature to the atom as a whole and then to atoms in contact with the atom receiving the neutral neutrino radiation. This energy could also be radiated out of the nucleus as gamma rays or possibly from electrons surrounding the nucleus as x-rays or even light.

Even though neutrino radiation interacts more than the present theory allows for, fewer neutrinos are observed coming from the sun. Why? This is because neutral radiation (a neutrino) is radiated in wavetrains. However, much of the time only parts of these wavetrains of neutrinos strike a neutron, and not enough energy is absorbed to cause the decay of the deuterium or chlorine atom. The present particle-like theory of neutrinos is like an on or off switch that does not make allowances for partial absorption. Therefore, the actual count is lower. The neutrino radiation created in the sun is also absorbed by the sun more than the present theory allows for. Since neutrino radiation interacts with matter more than the present theory states, the actual background of neutrinos is approximately half what the Standard Model predicts. These are the major reasons for the apparent deficit of neutrinos from the sun. [38]

With this view of neutrino radiation being a major cause of radioactive decay, it would appear that radioactive decay rates could vary slightly throughout the universe. This view can also explain why neutrons are stable in a nucleus but decay quite readily outside the nucleus. The kinetic or heat energy of protons and neutrons in a nucleus is actually quite low. Neutral neutrino radiation is everywhere. A free neutron, like a free electron, is capable of radiating and absorbing radiation in a much wider range of frequencies. Neutrino radiation strikes the free neutron heating it to an unstable state to where the electron and proton separate and radiate away any excess energy as more neutral neutrino radiation. Neutrons in the nucleus can only absorb neutrino radiation in harmony with the frequency it is oscillating at, reflecting away other frequencies. Also, a neutron in a nucleus that absorbs neutrino radiation can transfer energy to the other particles in the nucleus, as well as to the electrons surrounding the nucleus.

An experiment that may be able to prove this view of neutral neutrino radiation would be to compare the half-life of neutrons on the night side of the earth with the half-life of neutrons on the day side. The half-life of the day side neutrons should be slightly, but noticeably, shorter on average.

The WPTL easily explains how neutrinos are created in an extremely wide range of energies. [39] They are not particles. They are waves radiated in short pulses from neutral particles at extremely high frequencies.

How EMR Is Radiated and What It Is like during Transmission

Scientist understand that only electrons capable of oscillating at the radiating frequency are capable of absorbing these same frequencies. Now why is it that an electron is only capable of radiating certain frequencies, for instance the frequencies between a higher and lower stable energy level of a particular electron in a certain element? [56] An atom is a very complicated machine. What is often referred to as electron orbits is misleading. [55] Electrons don't really orbit a nucleus except possibly in hydrogen and helium. In other elements it is better to view electrons as in a relatively fixed position in relationship to the other electrons in this atom.

From the perspective of quantum mechanics, an electron's energy level is an accurate way of stating the electrons' distance from a nucleus. The difference is that in the WPTL, the electrons do not orbit the nucleus; [55, 56 and57] they retain relatively fixed positions in relationship to other electrons in all elements. Electrons in atoms contain linear and angular momentum; they oscillate inside its own electromagnetic field. These fields are not isolated from other electrons or the nucleus. So the frequency of each electron's oscillation must be in harmony with other electrons and the nucleus. This is why an electron, when given extra energy, radiates away this energy to return to a harmonizing state with the rest of the particles in an atom. [56] This view makes it quite easy to visualize why electrons don't radiate away all their linear momentum (heat) and fall into the nucleus. Electrons in the innermost shell and the nucleus are in a harmonic-like dance that does not have a lower harmonizing energy level. In a way, one could say the two innermost electrons in an atom keep each other from falling into the nucleus by their mutual repulsion in a harmonic dance with each other.

In many circumstances, EMR radiated from solids would be radiated in one general direction in long series of superwaves that could be described as wavetrains. Wavetrains also contribute to the photon-like nature of light. An interesting experiment to conduct would be to do the gold foil experiment with the EMR radiating from an extremely fast rotating source (high rpm). If light radiated from solids does have this wavetrain characteristic, the threshold of the frequency ejecting electrons would likely rise. A wavetrain is about a billion individual waves in each superwave radiated in a single direction by the tens of thousands to perhaps a million superwaves.

Just view a radiating electron as a billion machine guns packed as tightly together as possible and all firing in unison. In this theory, that's what is happening at the smallest level possible. These virtual or basic particles of light need nothing to travel through to get to the other side of the universe. But this is not possible even for these smallest of particles of light, they collide with each other.

The Alternating Spins in Waves of Basic Particles

The following is an assumption and may not be the only way an electron radiates away parts of the electromagnetic field surrounding it.

Since these basic particles travel at the velocity of light, one might ask how they can also spin. Wouldn't the rotating momentum actually be traveling in excess of the speed of light half the time? Yes, and this would be a problem. So they cannot actually be spinning when they are part of electromagnetic radiation. They can only be carrying a propensity to spin right or left. The total linear and angular momentum is always equal to 1/2h. In WPTL, at the speed of light, the basic particles carry only linear momentum and at total rest (absolute zero) only angular momentum. But when these particles are absorbed, much of this linear momentum is changed into angular momentum. The basic particles become part of the electromagnetic field surrounding the electron. They now contain linear and angular momentum equal to half a Planck unit.

The linear momentum is directly related to the heat and the electric field; the angular momentum creates the magnetic field of the electron. At absolute zero, the linear momentum of these basic particles surrounding the electron is zero and the angular momentum is, at its maximum, 1/2h. This may explain why magnetic fields are at their strongest at lower temperatures. It also explains why electrical resistance drops to zero. For with no heat (linear momentum), the electrons would flow without interference. You could say that the electromagnetic fields surrounding the electrons become closed loops with no place for magnetic fields from other electrons to become entangled with, and so there is no resistance. This occurs with double electron nuclei (Cooper's pairs) at slightly higher temperatures. It may be possible to increase the superconductivity of electricity to even higher temperatures with more electrons in an electron nucleus. (Electron nucleus, electrons fused inside their common or combined electromagnetic field is a state that can only exist at extremely low temperatures.)

This may also explain why a magnetic field is expelled from magnetic material when the temperature approaches absolute zero. As the temperature approaches absolute zero, more of the linear momentum contained in these basic particles is turned into angular momentum. This angular momentum creates a stronger magnetic field, but at the same time, the same spin of the magnetic fields comes into contact with each other and at a critical temperature, the field is expelled. (Like spins repel each other; more on this in chapter two.)

As mentioned earlier, a radiating electron actually radiates away some of the electromagnetic field surrounding it. The field surrounding the electron (E) is composed of basic particles with the same spin and angular momentum in the same phase, like this: ((((E)))). When the electron is at a high enough energy level to accelerate particles [>>] to the velocity of light, the first particles with spin > slam into the next layer of particles. Since the first and the second layer have the same spin, they exchange angular momentum and reverse spin. The second layer then slams into the third layer. Since the second and third layers now have opposite spins, they exchange linear momentum only. This means the third and fourth layer have the same spin and must exchange angular momentum etc, etc ...

This is why EMR is radiated and absorbed with alternating electromagnetic fields. While it is not known how many layers the radiation must travel through, it makes no difference. It always alternates.

This view of radiation consisting of waves of the smallest particles possible explains perfectly why the speed of light cannot be added to or subtracted from. Each half-wave particle is radiated and absorbed in the shortest time unit nature has to operate by.

Every oscillation of a radiating electron reduces the kinetic energy of the electron, which reduces its temperature. Heat and kinetic energy are interchangeable. This may not have much significance now, but later we will show that this is where a solar sail gains energy from by reflecting light. It does not come from a change in the energy or momentum of a photon or the change in the frequency of the reflected light.

If EMR is waves of particles with alternating spins in each half-wave, we should be able to combine light and increase the frequency. For example, if we combine the polarized laser light waves in the three lines below, we get the higher frequency in the fourth line.

Notice that when the first three rows are blended together, the frequency in the fourth row is tripled while still maintaining its alternating spins and electromagnetic fields. In theory, it should be possible to combine multiple polarized laser beams in odd numbers by blending slightly out-of-phase beams.

Solar Sailing

In the photon theory of light, it is the change in the momentum and energy (frequency) of the reflected photon that gives the solar sail its push. [11] How can this push come from a change in frequency when the reflected photon, at the start of this solar sailing adventure, has the same frequency as the photon striking the sail? There would be zero transfer of energy at the start of this type of solar sailing adventure.

Assuming the photons accelerated a solar sail to half the velocity of light, the reflected photons would be red-shifted to half their original frequency. One might think that half of the photons energy is used to accelerate the solar sail, but actually, the energy transfer is now half of zero. The light striking a solar sail moving away from the sun at 1/2c is red-shifted before it strikes the sail. In reality, only half as many light waves are striking the sail every second as it moves away from the sun at 1/2c. It is universally accepted, if one is traveling away from a light source at 1/2c, the red-shifting takes place before you see the light. An observer would see radiated blue light as red light. So where does the kinetic energy a solar sail gains from reflecting light come from if it doesn't come from a change in light frequency or the change in energy contained by a photon?

It comes from the heat in the solar sail. Just view the heat in the sail as atoms and electrons oscillating inside their own electromagnetic fields. When a light wave strikes this electromagnetic field that can't be absorbed, it is reflected. Two units of momentum are gained by this electromagnetic field in the direction of the incoming light wave. The first stops the light wave and the second reemits it in the other direction. This is no problem for the laws of conservation for momentum but this creates a violation for the law of conservation of energy. [8] Here is why. The reflected photon contains the exact amount of energy it struck the solar sail with. This energy cannot come from the light wave, because a light wave contains one Planck unit of momentum and energy (two half units), the smallest unit found in nature. It cannot come from the red-shifted reflected light, because every light wave striking the sail is reflected. So the light reflected from the solar sail is carrying away exactly the same amount of energy it struck the sail with. There is only one other place this energy could come from: the heat the electrons and atoms in a solar sail already contain.

One might ask, if the kinetic energy gained by a solar sail comes from the heat the solar sail contains, reducing its temperature, why don't mirrors feel cold as ice? When a mirror is fixed relative to the light source, it cannot gain kinetic energy, so all the potential kinetic energy gained by the mirror is recycled backward by the heat contained in the mirror and carried back away by the reflected light, leaving no net loss in the temperature of the mirror. If the light were to be absorbed, the temperature would increase.

One way of looking at the heat-kinetic energy interchange is to imagine a person catching and throwing a baseball on roller skates. On a flat surface, when he catches the ball, the kinetic energy from the ball pushes him backward. When he throws (reflects) the ball, it pushes him backwards a little faster. Assume the baseball is returned (reflected) with the same kinetic energy it contained before the ballplayer caught it. The ballplayer now rolling backwards contains kinetic energy and momentum of his own. How can this be? It appears he created energy by catching and throwing the ball back. This is not permitted. Obviously, the only answer is the kinetic energy the ballplayer now contains came from his own actions.

Now have this ballplayer with roller skates stand at the bottom of a dish-shaped hole. When he catches the ball there he is pushed up the backside of the dish. If he throws the ball as he rolls back to the bottom, he stops at the bottom exactly where he started. It is the same for electrons and atoms reflecting EMR. When fixed, they recoil, and upon reflecting the light, they end up exactly where they started. If the electrons and atoms are free floating, they recoil upon catching and reflecting the light. These electrons and atoms now have kinetic energy of their own. The question is where do they get this kinetic energy from? It cannot come from the reflected light, because the reflected light leaves with the same energy as it struck the electrons. The only possible source is from within. It comes from the heat contained by the electrons and atoms reflecting the light.

The present theory of using reflected light to power a solar sail is in conflict with the conservation of energy law. [8] We have already shown the red-shifting, if any, takes place before the light strikes a solar sail. Therefore, the photon is reflected containing the same amount kinetic energy as when it strikes the solar sail. So by reflecting the photon off the solar sail, we have increased the amount of energy in this closed system. So again where did the kinetic energy gained by the solar sail come from? Under the present photon theory of light, solar sailing would not work. [12]

Let's apply this view to the reflected laser light used to cool atoms into a Bose-Einstein Condensate state. [14 and 60] This laser light is reflected off atoms trapped by a magnetic field. The present explanation is that the light photons are reflected off the atoms at a higher frequency (blue-shifted) and this is what carries away the heat energy cooling the atoms. How can a reflected photon gain frequency other than by being reflected by an atom moving toward the source of the laser light? A light wave that strikes an atom can have only one of two fates; it can be reflected or absorbed. If a wave is absorbed it would add heat and be self-defeating. If the wave is reflected, it cannot carry away any more heat energy than it struck the atom with, one Planck's constant worth.

In the WPTL, some of the original laser light is red-shifted and some blue-shifted upon being reflected and plays no direct part in cooling the atoms. The reflected laser light turns the heat into kinetic energy and kinetic energy into heat in the atoms confined by the magnetic field. Let us first examine this process in two dimensions for simplicity, with atoms moving toward the laser light source. When these atoms reflect light, they are slowed down, which turns their kinetic energy into heat. This slowing process increases the pressures on these atoms, which causes an increase in their temperature. The increase in temperature (heat) is then radiated away leaving cooler atoms with less kinetic energy. With atoms moving away from the light source, the reflected light turns heat in these atoms into kinetic energy. These atoms then have more linear kinetic energy and are able to escape the magnetic confinement.

Let us now examine the laser cooling process in three dimensions with a laser light source from six sides. By repeating the processes mentioned above from all six sides it is easy to see that what the laser light really does is compress these atoms, which turns the atoms' kinetic energy into heat. The atoms then radiate away this increase in temperature leaving cooler less kinetically energetic atoms behind. The heat is driven out similar to the way a compressor drives heat out of a gas being compressed. In the case of laser cooling though, the heat is radiated away from the laser compressed atoms cooling them. Repeat this process many times and the remaining atoms are very cold, close to absolute zero.

It is at the electron energy levels in atoms that heat and kinetic energy are interchangeable. A similar thing happens in the Compton Effect. [13] The kinetic energy gained by the electrons comes from the heat they contain. Each x-ray half-wave striking the electron converts a little of the heat the electron contains into kinetic energy in the direction of the succeeding half-waves of x-rays. This increases the time before the second half x-ray wave collides with the electron. The red-shifting occurs before each of the succeeding half x-ray waves strike the electron. [58]

The energy contained in a wavetrain (photon) of reflected x-rays is the same before and after it is reflected. It is now spread over a longer period of time. The reflected x-ray has the same number of waves and particles as before, and is traveling at the same velocity. None of the kinetic energy gained by the electron comes from the wavetrain (photon) of scattered x-rays in the Compton Effect.

The kinetic energy gained by the electron in the Compton Effect is independent of the change in the frequency of the reflected light. [58] Therefore, the kinetic energy gained by the electron in the Compton Effect must come from another source. The only other source is the heat the electron already contained.

There is an interesting aspect of the Compton Effect. Why are the x-ray waves and electrons scattered at a ninety degree angle? It is believed in the WPTL that this is proof of the point-like nature of the particles that make up the EMR striking the electron. It is not proof of the point-like size of the electron. Therefore, electrons are not point-like particles. Individual half light wave particles are.

Refraction in the Waves of Particles Theory of Light

Snell's law and the Huygens' Principle describe reflection and refraction perfectly; the WPTL view of refraction does not change any of these principles or laws. The purpose of the following is to explain why and how this occurs.

When particles (1/2h) of light do collide, an interchange of linear and angular momentum can occur. This can create changes in velocity and direction of EMR. Collisions between these basic particles are the origin of refraction. You have probably seen refraction described as a row of soldiers marching off a hard surfaced road at an angle into sand. Since sand is hard to walk through, the first soldier to march into the sand slows down while the others are still marching at their original speed. This creates a bend in the line of soldiers. When the second soldier enters the sand, a new line is created with the first soldier that entered the sand at an angle to the other soldiers. As the soldiers march into the sand one by one, the original line disappears, and the new line forms at an angle to the original line of soldiers. This is a good way to describe how a beam of light is refracted as it travels into a more dense material at an angle. Just the opposite happens when the soldiers march from the sand at an angle onto a hard surfaced road. The same thing happens when a beam of light exits a dense material into a less dense material. But this description does not explain why light travels slower in denser materials. You need the WPTL to explain this.

A beam of light is composed of an astronomical number of basic particles which carry 1/2h of energy per particle from one place to another. As these basic particles enter a denser material, say from air into glass, they immediately collide with more basic particles that create the electromagnetic fields surrounding all the electrons and protons in an atom. Therefore, the basic particles composed of the light beam collide with more basic particle in this more condensed electromagnetic field. This means there will be more collisions between basic particles with the same spin. Collisions between basic particles with the same spin take an extra Planck unit of time to pass through each other. (This increase in time will be explained soon.) This slows down the first light particles from the beam that enter this denser material. Now just like the row of soldiers, the whole beam of light is bent as it enters this denser material at an angle. When the light beam exits, it is bent in the opposite direction for the opposite reason. The first light particles to exit return to their original velocity bending the light beam back in the opposite original direction.

These basic particles carry the smallest unit of force (momentum) possible, 1/2h at the velocity of light. The linear and angular momentum is exchanged in the smallest unit of time possible. Since exchanging linear momentum is a vector force in the same direction as the light particles are traveling, it takes no more time to exchange this force than it does for the particles to travel this distance. But exchanging linear and angular momentum takes an extra unit of time to exchange between particles with the same spin.

To graphically see why, visualize two linear lines of force crossing each other in one unit of space-time. You can see it takes no more time than it would for these particles to travel this distance. The force is exchanged as if the particles travel right through each other or as if the particles never exchanged linear momentum. This takes no time. With collisions between particles with the same spin, there is a sideways force that must also be exchanged. Now the lines of force must move perpendicular to the direction the colliding particles are traveling. This takes an extra unit of time. These two colliding particles are stationary in their linear forward motion for one Planck unit of time when this angular momentum is exchanged. Remember, these basic particles carry only 1/2h of momentum, so when particles with the same spin propensity collide, this spin propensity must be exchanged. The linear momentum changes to angular momentum and then back to linear momentum. This is the origin of time, mass, and gravity as we know them in our universe. (More on this in chapter two.)

Explaining why a beam of light is bent when entering a denser material at an angle is the easy part. It is more difficult to explain how a prism separates light by color.

The only way white light can be separated into different colors by a prism is if different colors travel at a slightly different velocity through the prism. As mentioned earlier, electromagnetic fields are made up of basic particles. When light enters the glass, it immediately encounters more collisions that exchange linear and angular momentum between these basic particles. This is why EMR travels slower in denser mediums. But this does not explain why higher frequencies of light travel slightly slower than lower frequencies through a dense medium. The reason we know higher frequencies of light travel slightly slower is because they are refracted more than lower frequencies of light.

The electrons in atoms are always exerting control over their own electromagnetic field. When light enters an atom, it changes this field into an alternating electromagnetic field. The electrons in the atoms immediately try to restore the electromagnetic field to the way it was before the EMR entered. The electromagnetic field is made up of virtual or basic particles surrounding the electron that have the same spin and electric field. What happens is at lower frequencies, the half-wave particles encounter fewer particles in the electromagnetic field that they must exchange angular momentum with.

Here is why. As waves of light particles travel through a single particle, they would exchange angular moment with every particle. Three waves traveling through a single particle in an electron's electromagnetic field like this < > < > < > travel through this >. The first half-wave exchanges angular momentum and reverses spin. All succeeding half-wave particles must also exchange angular momentum. It makes no difference whether you start with opposite spins or the same spin on the first collision. This is because if the first half-wave does not exchange angular momentum, the second half of the wave will, and all succeeding half-waves will exchange angular momentum.

While this is happening, the electron is trying to regain control of its electromagnetic field and reverse the spins to their original state >. At lower frequencies of light, there is more time for this to happen. So there are more instances of succeeding half-waves that do not exchange angular momentum, because the very next half-wave will now have the opposite spin of the particle in the electron's electromagnetic field. This leads to a slightly faster velocity of lower frequencies of light through this medium than higher frequencies of light. At higher frequencies, just the opposite happens. With less time for the electron to restore its electromagnetic field's particles to their original spin state >, more of the succeeding half-waves must also exchange angular momentum, which slows the light's velocity through this medium.

It is this increase of exchanges of angular momentum between these basic particles that causes higher frequencies of light to travel slower through a denser medium. It is the electron's constant effort to regain control of its own electromagnetic field that causes lower frequencies to travel slightly faster than higher frequencies. This is a prism at work, separating the light by colors.

From the preceding description of refraction, it appears negative refraction would be impossible.[9] There have been several claims made of negative refraction in the last few years but it is highly questionable whether negative refraction has been achieved. All claims have one thing in common: it appears all have the light travel through some sort of crystalline material or tubes. [10] These negatively refracting materials appear to be nothing more than waveguides for higher frequencies of EMR.

Another Effect of the Refraction of Light

There is a second effect of refraction that is only noticeable at the half-wave level. This type of refraction occurs at a unit of space-time smaller than the smallest units of space-time, a Planck length. [29] This sounds like a contradiction. These basic particles may be the smallest units of space and time, but in theoretically empty space, smaller measurements are possible. The universe doesn't operate in these smaller measurements of space and time. For example, suppose two of these basic particles of light traveling in opposite direction collide head on. Suppose they have the same spin. They exchange spins, go on their merry ways, and there is no problem. But what if the collision is not direct, but less than a half Planck unit of distance off from being direct? The WPTL proposes that they would be refracted into each other as if it were a direct hit.

In the WPTL, it is the collision between light particles traveling in opposite directions that is the source of the red-shift observed from the distant parts of our universe.

We only need to deal with single waves, because little light, if any, would arrive across these vast distances as superwaves, let alone wavetrains. As mentioned earlier, all succeeding half-waves are radiated and absorbed with alternating spins in electromagnetic fields. There are exceptions to this rule. Between the points of radiation and absorption, the half-waves often have spins that are the same.

How is this possible? It is possible because of the half-wave particle refraction. In other words, light particles traveling in a nearly perfect opposite directions are refracted into each other. The first half-wave particle refracts some particles of light into its path from light waves traveling from the opposite direction. These particles may also travel through the second half-wave particle and, in addition, to any particles the second half-wave may refract into itself. Since the second half-wave travels through slightly more particles of light from the opposite direction, there is a 50% chance these collisions will exchange angular momentum. This increases the distance between the first and second particles in a wave by one Planck unit of space-time, its frequency is reduced or red-shifted. Over the billions of years these light waves have been traveling, it adds up to the source of the red-shift observed from the distant parts of the universe.

A test which may be able to prove this origin of red-shift could be to have a single frequency of laser light travel in opposite directions in a fiber optic cable for as great a distance as possible. Use continuous beams, not pulses, and then recombine the beams after traveling this distance with light from the original source. Compare the frequencies for any red-shifting. A test over fifty kilometers may not show a measurable red-shift, but the chromatic dispersion of light clearly shows up because some of the multitude of waves travel through less dense electromagnetic fields in the atoms just by chance. [15]

The Waves of Particles Theory and the Universe

This view as to the origin of red-shift observed from distant galaxies would change many things we now accept as fact. The Big Bang Theory would no longer be needed to explain the expanding universe, because the universe would not be expanding. [16] The expansion would just be an optical illusion. But the arguments against this will be many. One argument will be that it violates the law of the conservation of energy and momentum. [8] It does not violate this law, because there is a one-for-one exchange. As the frequency slowly red-shifts over billions of years, the intensity of the reduced frequency goes up in direct proportion. In any given second of space-time, the number of half-waves is the same. So if light from a distant galaxy is red-shifted by a factor of one, meaning its frequency is reduced by half, the intensity of the red-shifted light from this galaxy is two times as bright or intense. Light red-shifted by a factor of two would be three times as intense, and so on. The number of waves or particles of light in any given unit of space-time doesn't change. Energy and momentum are conserved. If EMR is viewed as millions of waves of particles traveling parallel to each other and out of sequence in a beam, it is easy to see how the intensity of the beam would increase as the light is red-shifted. As the waves are red-shifted, they simply overlap waves traveling in parallel. Therefore, in any given second of space-time, the number of waves remains the same, but at a lower frequency. That is what increases the intensity of the lower frequency's beam.

An equal distance of space-time still contains the same number of half-waves or particles of light. The waves simply start to overlap adjacent waves.

In this view, distant supernovae should appear brighter than their distance allows for. The intensity of the red-shifted light has been increased in direct proportion to the red-shift factor. But they appear dimmer. [20] Why? The very measurement of distances would also be changed for distant galaxies. For example, if a red-shift of one is equal to eight billion light-years, then a red-shift of five would be in direct proportion, or forty billion light-years. In short, the supernovae are farther away than the Big Bang Theory allows for. Some more recent findings of very distant supernovae show a return to brightening, which is a little confusing. [18 and 21] But remember, in this theory as the light is red-shifted, the intensity of the light increases in direct proportion. It is this apparent contradiction in brightness to distance of observed supernovae that required the acceleration to be added to the expansion of the universe. [19, 25 and 26]

So if a Big Bang did not create the universe, what did? And what is the shape of the universe? Neither question can be answered at this time. The best that can be said is that the universe is apparently infinite in size and had no beginning. Nor will there ever be an end. This is not the same as saying our solar system and galaxies have no beginning or end. They do. We live in an endless recycling universe, a universe where galaxies are born and eventually disappear. The life cycle of galaxies is probably in the trillions of years. Matter and energy are endlessly changing into each other. [17]

What about the Cosmic Background Radiation (CBR)? [22] It fits in with an endless infinite universe much better than a created Big Bang universe. In fact, it does not even fit a created Big Bang universe. That is one of the reasons why the concept of inflation had to be added to the Big Bang theory. [27] Inflation is the concept that the universe expanded at millions, if not billions, of times the velocity of light in the first few seconds after the Big Bang. The concept of inflation was needed to explain the near uniformity of the CBR from every direction astronomers looked into the universe. They either had to add inflation or claim we happen to be at the center of the universe. They weren't about to make that claim again. The ironic thing is it would appear we are at the center of the universe. In an infinite universe, everywhere is the center.

Nor does the fact the universe is apparently flat support the Big Bang Theory. It supports an infinite universe. [24] This is another reason the concept of inflation had to be added to the Big Bang Theory. A flat universe means that the light reaching us from every direction is uniform and apparently arrives in a straight line from every direction. This also fits in with an infinite, ageless universe without any modifications needed. The red-shifted light we see looking out into an infinite recycling universe is just an optical illusion.

Yes, but there are other problems with an infinite universe. Wouldn't the universe be bright instead of dark? No, it still would be dark because EMR is red-shifted down to the radio wave range over billions and billions of light-years. What about gravity? Wouldn't it become infinite? No again. In chapter three, we will see that these same basic particles use by EMR are the origin of gravity. The easiest way to view this is to compare it to the electromagnetic force. In theory, the electromagnetic force goes on to infinity. In reality, it only goes as far as another electromagnetic field that is stronger than the first field in that part of space-time. From there out farther, the electromagnetic field's influence is reduced slightly faster than the inverse square rule calls for. In theory, ocean waves would go on forever, but not in the real world. The reason is that there are only so many molecules of water and basic particles to transmit these forces. A molecule cannot go up and down at the same time. Neither can basic particles transmit two quanta of momentum. This is the basic nature of a quantum universe, only a finite amount of momentum and energy, and therefore information can be sustained in any given area. Some information is continually being lost forever in any given area of space-time.

What about entropy? [28] Wouldn't the universe unwind and eventually become devoid of energy and life? Unlike mankind, nature can recycle everything given enough time. Black holes and neutron stars are no problem. Stephen Hawking figured out mathematically how black holes can eventually evaporate. Neutron stars can also evaporate. They are a ball of neutrons that absorb most neutrino radiation that strikes them. This heats them up neutron by neutron. These neutrons, just like a free roaming neutron, decay into an electron and a proton. Now what do you have? You have a hydrogen atom to start the whole process over again. Energy and matter are interchangeable. It is not a one-way street, matter changing to energy. Energy can also change into matter. An example is gamma rays creating electrons and positrons. [4]

This new theory of light may never be directly provable. It may only be accepted by a process of elimination. When everything else fails, you accept that which fits reality the best.

The next chapter, "The Origin of Matter and the Electromagnetic Force," is about the origins of the forces and particles around us. This chapter also will look at the micro- and macro-world in strange new ways. For example, there are no attractive forces in nature and opposite charges are not attracted to each other.

2

The Origin of Matter and the Electromagnetic Force

In the Beginning

The beginning question, where everything came from, cannot be answered at this time. Furthermore, it may be unanswerable. So let's just imagine how things may have been at an imaginary starting point in a recycling, everlasting universe. Assume that empty space is teeming with zero-fluxing energy at the smallest dimension possible, a Planck length, and the smallest unit of energy and momentum, a half Planck unit. Imagine these Planck units are just the natural resonance of empty space. It is like a wave that resonates down to a level to where it resonates with itself and splits into two half units. These two half units travel away from each other in opposite directions at the velocity of light. The sum of the two ½ Planck units of momentum, one positive and one negative, equal zero. These half Planck units are the very same units (particles) that the wave of particles theory of light (WPTL) is based on. This creation and destruction of ½ units of Planck momentum is going on everywhere in the universe in numbers unimaginable even in atomic size volumes of space. They are the origin of dark energy.

Imagine these half-waves [< or >] as a positive and a negative momentum equaling zero. They are mirror images of each other, and they are reversible. So when they resonate off each other, it is not positive or negative energy, it is just energy. One could say there is no positive and negative energy in the universe, just energy.

Since the creation of these virtual, or basic particles, is just the nature of empty space and they zip around in unimaginable numbers, they will collide with each other. In the WPTL, there are two different ways these collisions can take place, as an exchange of linear momentum only with opposite spins or with the exchange of linear and angular momentum with the same spins. These basic particles can be viewed as strings traveling at the velocity of light, with the string (force or momentum) to the right or left of center. These strings do not rotate when traveling at the speed of light. But since the force carried is to the right or left of center, they have a propensity to spin right or left. We will view these basic particles with spin or vibrating strings, depending on the circumstances, because sometimes it is easier to visualize what is happening with spin and at other times with vibration.

This theory assumes that there is a smallest Planck unit of time and length. While it cannot be stated with any degree of certainty the value of these units, there must be the same numerical quantity of both in the distance light travels in one second. It is widely believed that there are 10^{43} units of time in one second.

Therefore, there would be 10^{43} units of space in the distance light travels in one second, 300,000 kilometers. [29]

When these particles collide, it is like two spinning tops colliding. If they have the same spin, there is an exchange of linear and angular momentum. When spinning opposite, only linear momentum is exchanged. As mentioned earlier these basic particles only have a half-unit of Planck momentum and travel at the speed of light. So when two particles with opposite spins collide, they exchange linear momentum in the same amount of time as it would be to travel this distance at the speed of light. It is as if they travel right through each other, and it may well be they do because there are two particles still traveling in the same directions, at the same velocity, and with the same spins as before the collision. It's like nothing ever happened.

A Universe Blossoms

While nothing seems to happen when two of these basic particles with opposite spins collide, a whole universe arises from the collision of these same basic particles with the same spin. The universe arises from nothing, not at the Big Bang level, but at the smallest level possible and everywhere. When these basic particles with the same spin collide, angular momentum must be exchanged, a force perpendicular to the velocity of these particles. Therefore, it takes one extra unit of Planck time for these particles to exchange their linear and angular momentum.

This makes for some interesting results:

(1) The extra unit of Planck time is the origin of time as we know it. From this origin we can see that time is measurement of change. Therefore, it can only have a positive value. One could give a positive and a negative value to the different particles, but it is meaningless because it takes the same amount of time for the angular momentum to flow either way, and a difference cannot be measured. A measurement of change is always a positive value. Change is change; time cannot flow backward. All changes in the universe are the result of the cumulative effects from the exchange of angular momentum in Planck length-sized particles.

(2) The spin exchange of these colliding particles is the origin of mass as we know it. In the unit of time it takes these basic particles to exchange angular momentum, they are stationary. They have no apparent linear movement. It would take energy to move them in this one quanta second of time. They appear to have inertia or mass.

(3) Since it takes an extra unit of time for these two particles to exchange their angular momentum, they will be one Planck unit of length closer to each other

after the collision. These two particles turned one unit of Planck time at the velocity of light into one unit of mass and are one Planck unit of space-time closer together. It would now take energy to place them as far apart as they were before they collided and exchanged angular momentum. This is the origin of gravity. So gravitons are the very same particles as the particles in electromagnetic radiation (EMR). They are just a different aspect, the exchange of angular momentum. This explains why gravity is such a weak force. In a way it is not even a force, it is the momentary absence of linear momentum, a lower dark energy level.

There would be nothing without these colliding basic particles that exchange angular momentum: no time, no length, no gravity, no particles, and not even any forces. The universe would not exist; in its place would appear to be nothing.

Matter: Electrons and Protons

It can be visualized how these basic particles create matter, but there is a bit of a chicken-and-egg problem. We know EMR can create an electron and a positron. Assume extremely powerful gamma rays can also create protons and antiprotons. The problem is one cannot have gamma rays without the particles. So which came first? That just cannot be answered at this time, nor can the question of what happened to all the antimatter. The most likely scenario is that half the universe is antimatter with galaxies being mostly a single type of matter, even galaxy cluster are probably matter or antimatter.

One could speculate that the basic particles that exchange spin start to condense in every increasing density. This may be going on right now in some of the great voids in the universe or at the center of galaxies. If something like this is happening, these basic particles may collect into the densities of black holes. They would be so tightly packed that they would have no forward movement or linear momentum. They would only be exchanging angular momentum with each other.

In the WPTL, an electron and a proton are made of a finite stable number of basic particles that exchange angular momentum into infinity or until they meet their antiparticle. But why certain sizes, like the stable size of an electron and a proton? That cannot be answered at this time either. But one can speculate as to why there is a difference in the size of an electron and a proton. It must be the smallest stable sizes. Anything less than these differences would not be stable. Negative electrons, in numbers (1836), might be able to annihilate a proton. Or possibly, it is a stable internal oscillating state of the basic particles in electrons and protons.

To get an idea of the size of a basic particle, compare it to the electron. The electron has the energy equivalent in mass of about 511,000 eV. If a single Planck's constant is equal to 10^{-16} of 1eV, then a single electron, which is considered a point particle with no dimension, would be composed of about 10^{22} of these basic particles. This gives some idea about how tiny the particles are that this theory is based on.

Gamma rays can create matter—electrons, and positrons. The WPTL can explain this. Gamma rays striking a nucleus are stopped and turned into angular momentum. Since every other superwave has an opposite spin, when they collide with the nucleus, they form two separate "piles" with opposite spins, one to the right and one to the left. Now the basic particles in each pile all have the same spin. As stated before, particles with the same spin exchange angular momentum. They appear to have mass or inertia. This spin exchange will continue indefinitely until a pile meets an antipile. An electron and a positron are born. Protons and antiprotons may also be created this way by gamma rays with even higher frequencies. It may also be that matter, electrons, and protons are created in black holes. But this leaves the question where do the positive and negative charges of these particles come from?

The Origin of the Electrical Charge

The following explanation is a possible origin of the electrical charge. If particles with mass—the electron, proton, and their antiparticles—are made up of about 10^{22} to 10^{25} of these basic EMR particles that exchange angular momentum, it may be possible to offer an explanation. It has been explained where mass comes from, the exchange of angular momentum between these basic particles. With the electrical force, it may be easier to view these basic particles as vibrating strings or view the angular momentum contained in these basic particles as being in two time frames. It would take these basic particles two quanta seconds to complete one revolution, or cycle of vibrating, at the velocity of light. Consider the angular momentum of the basic particles in an electron or proton as facing out from the center of the particle, or inward toward the center, every other quanta second. While the angular momentum is facing inward for the electron, it is facing outward for the proton and vice versa.

This equal and opposite oscillation of the angular momentum of the basic particles in the proton and electron is what creates an equal and opposite standing wave in the basic particles surrounding a particle with mass. It is this standing wave that creates and transmits the electrical force. It is like there is an oscillating pressure surrounding "charged" particles with mass. The oscillating pressure, like in waves, has a positive and negative value. The standing charged wave pressure is inward toward the particle one quanta second and outward the next.

When two like particles with mass approach each other, the positive and negative pressure waves are in the same time frame. This increases the pressure backwards toward both particles, say electrons, and they repulse each other. When two oppositely charged particles, such as an electron and proton, approach each other, the standing waves in the basic particles between them being in opposite time frames cancel each other out. They neutralize each other's electric field between them. This reduces the pressure a particle with mass creates from the standing wave it created around itself. The back-pressure the charged particles "feel" is still the same on the other side. The electron and proton are not attracted to each other. They are repelled by their own now unbalanced electrical fields toward each other. In an atom, the electron's negative field is not uniform. It is this weaker spot toward the nucleus that confines the electron to the atom. Place trillions of atoms in two groups and give them opposite electrical charges and they are "attracted" toward each other in the same manner. The atoms oscillate toward each other by their own, now unbalanced, electrical fields.

Charged electrical fields must contain some heat to be either repelled or attracted toward each other. At absolute zero, the electrical force would appear to vanish. Charged particles must contain heat to oscillate toward or away from each other. The kinetic energy gained from this motion does not come from the heat these charged particles contain. It comes from an imbalance in the charged electrical fields upon their creation. Energy was stored in the creation of these electrical fields. That is where the kinetic energy gained by charged particles comes from when they are attracted toward each other.

When an electron is captured or falls to a lower energy level in an atom, energy in the form of radiation is released. This energy is released because the electron is now bound by a stronger charged field than it was farther from the nucleus. This increases the pressure, and therefore the frequency of oscillation is increased. The amount of heat contained by this electron is reduced as it radiates away some of this increase in frequency. (More later as to why this is.) The electron is now at a lower energy level. The energy radiated away would have to be replaced to free the electron or ionize the atom again.

In this theory, the electron is composed of two parts similar to an egg. The yolk is the particle, and the white surrounding the yolk is the electromagnetic field. The electron vibrates around inside the negative electromagnetic field created by the heat or stored kinetic energy it contains. It is the electromagnetic field that keeps the vibrating electron contained and oscillating. It is what is putting pressure back on the electron. It confines the electron, storing the energy received from EMR or neighboring electrons in the form of kinetic energy or heat.

The WPTL view of particles with mass as having two parts can explain the double slit experiment in which an electron displays the dual wave and particle properties.[54] With both slits open, the particle part goes through one slit, but some of the electromagnetic field surround the electron goes through both slits. When the electromagnetic field recombines on the other side, it creates interference and deflects the electron. The electron (particle) never goes through both slits at the same time; only part of the electromagnetic field surrounding it does. Electrons and protons are both a particle and a wave; only at absolute zero would the wave characteristic of these particles disappear. At lower temperatures, particles with mass are oscillating at lower frequencies, and the wave characteristic is quite noticeable.

It is this electromagnetic back-pressure, surrounding the charged particle, that enables two oppositely charged particles to be attracted to each other. As they approach each other, the opposite electromagnetic fields between them are neutralized. The electromagnetic pressure on the outside remains the same, and the oppositely charged particles are forced toward each other. The heat caused oscillation is now unbalanced, and the charged particles migrate toward each other. This is what holds atoms together. The electrons fall into lower stable energy levels, radiating away excess energy. There is nothing wrong with saying opposite electrical charges attract each other. The whole purpose of the preceding explanation is to explain how this could happen.

This view may also offer an explanation as to how lightning works. Charges in clouds may be low voltage differences over a large area. Say a cloud only has a ten voltage difference from the ground. The number of extra electrons in this cloud may be a huge number; that is the amperage potential may be large. The extra electrons in this cloud by their own mutual repulsion would collect on the outer surface or side of a cloud. The electromagnetic field created by these electrons would overlap each other. This creates an electromagnetic field that is capable of confining more electrons that are expelled into the electromagnetic field. The same thing happens in electric wires or in a Van de Graff generator's sphere that holds a powerful charge. [30]

The ground below this cloud has the opposite charge. High ground or a tall tree that does not have the smooth surface of a sphere would create a weak point on the surface of the electromagnetic field containing the low potential voltage in this cloud. From this point, the electromagnetic field surrounding the low voltage difference in the cloud has a point of discharging. As the cloud discharges, the extra electrons in the huge bubble surrounding the cloud all start heading in the direction of the discharge point. All this time they are picking up speed, or kinetic energy. As mentioned earlier, these electrons must contain heat to turn the electric charge into kinetic energy. This may explain why lightning is more prevalent during warm weather. These electrons are driven by their own collapsing electromagnetic field that they created. As the electrons gain momentum flowing toward the discharge point, the voltage increases. The increase in voltage comes from the increasing density of electrons as they collect at the discharge point. This ionizes the air and creates a good conductor of electrons. This reduces the charge back-pressure of the collapsing bubble. Literally trillions of extra electrons from a low voltage source escape through this point, transforming a large, low voltage source into a small, high voltage discharge point: lightning.

The preceding explanation as to the source of the electrical charge does create a difficulty though. If the electrical charge originates from a division of time into two frames, wouldn't temperature or the oscillating fields throw the two time frames out of sequence? It would. From Einstein's theory of relativity, we know that velocity affects time. Time is altered by the same value or multiples for all the electrons and protons in atoms and electromagnetic fields. The time frames for the protons and electrons that have a very different oscillating velocity at the same temperature do not really have to be in the same frequency, only harmonizing time frames or resonating frequencies. This is the origin of the electron's energy levels in the atom. The electron radiates away excess heat to achieve the next lowest harmonizing energy level with the other electrons in this particular atom. This places the electron in the next lower harmonizing frequency with the other electrons and nuclei in this element. The positive and negative electrical time frames also change during this drop to a lower harmonizing frequency. They are self adjusting to any temperature.

The role temperature plays in the electrical and magnetic force may explain why electrical resistance drops as temperature drops, while the magnetic force increases with the drop in temperature. Since the oscillating frequency of the electron decreases with a decrease in temperature, there would be less times frames for the basic particles creating these fields to be out of step with the two time frames in a charged field.

The Origin of the Magnetic Force

The magnetic force is created by the spin of the basic particles surrounding charged particles, such as protons and electrons. The basic particles tend to line up by spin in long chains called magnetic flux. This is the same flux that is radiated away at the velocity of light in the WPTL. What we end up with are long chains of basic particles with the same spin and with the angular momentum of the basic particles in these chains rotating in the same time frames. That is when the angular momentum of the basic particles is facing outward from the electron; it is facing inward for the proton and vice versa. We have an electromagnetic field.

Electrons and protons are in a way imprisoned by the very electromagnetic fields they create. They oscillate around inside this field. This is where kinetic energy in the form of heat is stored.

If the electrical force is only a repulsive force, certainly the magnetic force must be repulsive and attractive, right? No—not in this theory. In the WPTL, there are no attractive forces in nature. An attractive force by its very nature gives the appearance of something magical. There is no room for anything magical in this theory. Everything originates from the momentum (force) contained in basic particles.

As in the electrical force, a magnetic field creates pressure back on the particles with mass that create it. Like spins in the basic particles resist entering each others space; they repel each other. This is plain to see by the way iron filings orient themselves around the two poles of a bar magnet. The lines spread apart at both poles; they are repelling each other's like spins. What is more difficult to visualize is how two like poles approaching each other from opposite directions repel each other. It would appear they are now approaching each other with opposite spins and should not repel each other. Similar spins repel each other, just like in tops. Imagine that we cut a bar magnet in half length-wise (north to south pole) and rotate one half to face the other half from the opposite direction. Now these two like poles still repel each other, even though it would appear the basic particles in the magnetic fields of the two like poles now have opposite spins. Why? When we split the magnet in half and rotate one half to face the other from the opposite direction, the two fields now with what is an opposite spin still engage each other with the same spin. Now this is really appearing strange.

If one follows the lines of flux in the iron filings from two like magnetic poles facing each other, one can see why they engage each other with the same spin. All the magnetic lines of flux have the same spin originating from a single pole, so they repel each other. This is true even down to the level of electrons; the lines of force repel each other at both poles. To keep this simple; let's view it in two dimensions. Assume the magnetic flux lines had spin right before we cut the bar magnet in half. Then the original magnet's lines of flux that curve to the right and have a spin to the right, now have a spin downward. The lines that curve to the left and have the same spin to the right, now spin upward. Place the two like poles toward each other from the opposite direction. The rotated to the top magnet's magnetic flux lines now have spin left. The lines that curve to the left now have a spin upward and to the right downward, just like the other magnet. Draw this out on a piece of paper, and it is easy to see. It is the same spin of the basic particles that cause the two like magnetic poles to repel each other. The same magnetic poles' opposite spins engage each other down to the smallest magnetic level, two electrons with the same spins.

Opposite Magnetic Poles

How are opposite magnetic poles attracted to each other? Let's take a bar magnet and cut it in half as if to separate the north and south poles. We know that this is impossible because we will just end up with two shorter magnets, each with their own north and south poles. These two new smaller magnets would resist being pulled apart after the cut was made. Why? This is more difficult to explain. We know the magnetic flux lines all have the same spin inside this bar magnet before and after it is cut into two shorter magnets. It would appear that the two newly created poles facing each other should repel each other because they have the same spins. But we know they don't; they attract each other. One way of looking at this is to go back to the spinning tops. Tops with like spins repel each other, but this is only true when they approach each other from the sides. If they approach each other from the bottom or top with the same spin, they will spin together with no repulsion. If the tops approach each other with opposite spins from the top or bottom, they will repel each other. This is exactly why like poles with the same spin, when approaching each other from an opposite direction, engage with opposite polar spins. They repel each other's magnetic field. Likewise, opposite poles engage each other with the same polar spin and do not repel each other.

Another way to see why opposite magnetic poles attract each other is to pull the two bar magnets we created from the one we cut in half away from each other. Now let's examine the spins of the two opposite poles magnetic fields. Just as when two poles with opposite spins engage with the same spin, here the same spins engage with the opposite spins. Again in two dimensions, follow the spin in the flux lines. The magnetic field lines of flux repel each other even down to the level of electrons. So the flux lines repel each other at both poles. Let's assume that the magnetic field's of these two unlike poles has spin left. The lines of flux that curve to the left from the bottom magnet now have a downward spin, and the flux to the right has an upward spin. The upper magnet also has spin left, which, when it curves to the left, has a spin upward, and when it curves to the right, has a spin downward. So the two opposite poles engage each other with opposite spins, and do not repel each other. The spin at the poles being the same also do not repel each other. The magnetic lines of flux link up from the opposite poles just as they were before we cut the magnet in two.

Now this may explain why they don't repel each other, but it does not explain why they are attracted to each other. They are not attracted to each other; they are repelled toward each other by their own opposite poles' magnetic fields. As explained earlier, charged particles are contained by the back pressure of the very charged field they have created. It is the same with a magnetic field, only this is at a right angle to the electrical field. The magnetic field places a pressure at the two poles of the charged particles creating this magnetic field. The two opposite magnetic poles neutralize this pressure, and back-pressure from the other poles pushes the two magnets toward each other. Just like in the electrical force, these bar magnets must contain some heat to be attracted toward or pushed away from each other.

It is heat as stored kinetic energy in an electromagnetic field that enables charged particles or magnetic fields to release the stored potential electromagnetic energy. At absolute zero, the potential electrical or magnetic energy is locked up; it is unusable. With even a little heat, the electron and protons in atoms are oscillating. This oscillation is necessary for the electromagnetic fields to engage each other, thereby either repelling or attracting each other. When stored potential electrical or magnetic energy is released, the temperature needed for the release is not affected in any way.

The reason heat is required for a charged or magnetic field to engage is because the electromagnetic field surrounding the charges or poles would not be oscillating at absolute zero, and therefore they will not engage. If these fields do not engage, there is no way for the charges or poles to repel each other either away from or toward each other. When the charge or magnetic field engages, it creates an imbalance in the electromagnetic field surrounding the charged particles. There is a surplus or deficit of a positive or negative charge between the charged particles now. This is where the heat contained by these charge particles comes into play. The heat (oscillation) becomes unbalanced and spends more time going in one direction as opposed to the opposite direction when electrical or magnetic fields engage. This is very similar to the way a Mexican jumping bean turns internal motion into linear motion.

When an electrical or magnetic field is created, it is like a valley and a hill are created. The opposite charges or poles are these valleys and hills. When opposite charges and poles engage, they seek a common equal neutral ground. When the same charges and poles engage, the hills or valleys are enlarged. It takes energy to create this larger charge or magnetic field.

Have you ever tried to force the like poles of a strong bar magnet together? If you have, one thing you noticed is that it is impossible to keep the center of the like poles on each other. It seems to continually force your hands off to one side or the other. This is because the electrons creating the magnetic field are in constant random motion with each other caused by the heat they contain. The magnetic flux lines also have a random wave-like motion down to absolute zero. This makes it impossible to place two like magnetic fields exactly in the center of each other's magnetic field. Just the opposite happens with opposite poles; they are centered with ease. With unlike poles the magnets are pushed together to the lowest possible energy level, which is as close to the center of these opposite poles as possible.

There can be no magnetic monopoles in the WPTL. You cannot have one end of a rotating object without another end.

The Neutron

In WPTL, a neutron is a composite particle made of an electron and a proton. The electron is in an extremely low kinetic energy level orbit around the proton. Both particles are inside their now combined electromagnetic fields. Their opposite electromagnetic fields create a new neutral field. As mentioned earlier, in the electromagnetic field surrounding the electron and proton, the basic particles form long chains with the same spin and with their angular momentum in the same time frame or phase. Now the electron and proton's electromagnetic fields have the opposite spin and charge. The basic particles cannot complete a revolution; instead, they alternate exchanging angular momentum. This exchange of angular momentum is where the extra mass of the combined electron and proton come from. Much of the energy contained in the electromagnetic fields turns into mass. The neutral field created around them is not able to interact with a positive or negative field. A neutron does have a very weak magnetic field surrounding it.

It is believed that the electron rotates around the proton, but the electron and proton also rotate around as a single unit creating a weak magnetic field. This is like the weak magnetic field an atom can create, which also has a neutral overall charge.

In chapter one, it was mentioned that neutral neutrino radiation had no mass at the velocity of light, but that it had a rest mass. The reason it has a rest mass is because the basic particles in the two opposite electromagnetic fields exchange angular momentum, and it would take energy to accelerate them to the velocity of light. But once they are accelerated to the velocity of light, they no longer contain angular momentum, only linear momentum. Therefore, they no longer exchange angular momentum and don't contain mass as we know it, only linear momentum. Like EMR, neutrino radiation cannot travel at the speed of light and still rotate, because half the time the rotating string of these particles would be traveling faster than the velocity of light. This is forbidden. Nothing in nature happens faster than the speed of light.

Combine an electron ((E)) and a proton))P ((. This composite particle called the neutron would look something like this: ( )( )EP( )( ). The parentheses on both sides are the opposing electromagnetic fields; the E and P in the center represent the electron and proton. The Standard Model states that the neutron is an independent particle created out of three quarks. But close examination of the neutron offers little support for this view; it supports the composite particle view. A much simpler view is one with two fundamental particles—the electron and proton—that retain some of their individual identities. [59, 72 and 73]

The 1956 experiment showing parity violation may actually be evidence that a neutron in not a fundamental particle but a composite particle created from an electron and a proton. [31] In the following chapter, it will be explained why neutron decay on occasion violates parity.

Another piece of evidence that the neutron may be a composite particle is that there is a slightly positive center charge surrounded by a slightly negative charge. [59, 72 and 73] How can the quark theory explain this?

One would wonder why, when a proton in a nucleus changes into a neutron, it requires an electron if a neutron is not a composite particle created from an electron and a proton. When a proton changes into a neutron in the nucleus it is always accompanied by electron capture. [61] But also, when a proton changes into a neutron, a positron is emitted. This presents a problem. A positron is created from the decay of the proton into a neutron while at the same time the proton is capturing an electron to change into a neutron. If this is what really happens wouldn't we end up short two electrons? How is this possible? Aren't the positron and electron always created in pairs? The answer is in the next chapter.

When neutral neutrino radiation heats a free neutron, the electron and proton oscillate inside their neutral field in an ever-increasing fashion. Finally the electron slams against the proton with enough force to escape from their combined electromagnetic fields, decoupling the fields. Some of this heat energy is transferred into kinetic energy in the electron and proton and some is radiated away as more neutral neutrino radiation. Some of this kinetic energy also comes from the mass created from the angular momentum exchange stored as mass in the neutral electromagnetic field around the neutron.

The next chapter will explain the other three fundamental forces in nature. It will also explain exactly how gravity works. The explanation is both stranger and simpler than one might imagine.

3

The Forces of Nature

In the Wave of Particles Theory of Light (WPTL), there is only one force, the electromagnetic force that was covered in chapter two. What about the force of gravity and the strong and weak nuclear forces?

The Weak Nuclear Force

The weak nuclear force is not a force in this theory. Its source originates from the unseen and random neutral neutrino radiation explained in the first chapter. Radioactive particles do not spontaneously decay. They are struck and absorb neutral neutrino radiation placing the nucleus in a higher unstable state. That is what causes them to decay. Stable elements do not decay, because the natural background of neutral neutrino radiation striking them does not contain sufficient energy to overcome the strong force binding the nucleus together. It is simply radiated away from the nucleus as gamma rays or more neutral neutrino radiation. It can also be passed on to the rest of the atom in the form of heat and then simply radiated away. Some isotopes with extremely short lives never reach a low enough energy level to be considered stable for any period of time, even in the absence of neutrino radiation at any temperature above absolute zero. Beryllium 8 is an example of this type of isotope.

Some more rare forms of decay, like the positron emission mentioned earlier, can also be explained. The reason it is rare is because it takes a lot of neutrino radiation to accomplish this form of decay. There must be sufficient energy to create a positron and the electron. The electron combines with a proton forming a neutron, and the positron is ejected from the nucleus. The positron then collides with an electron, and both are annihilated turning into gamma rays. No extra electrons here. Not all the energy needs to come from the random neutrino radiation; some can come from the nucleus dropping to a lower energy level, which is a more stable state.

Radioactive decay, as in quantum mechanics, is completely random. There is absolutely no practical way to predetermine which radioactive atom in a group would decay next. One would have to know everything that has ever happened to every particle in the universe to know exactly when neutrino radiation of sufficient energy would strike a particular radioactive atom causing it to decay.

Neutrino radiation is not the only factor that can cause radioactive decay. High energy types of electromagnetic radiation (EMR), such as gamma rays, can also set a chain of events into motion that lead to radioactive decay. [70]

There are several facts that support the idea that neutral neutrino radiation is the major source of radioactive decay. Muon neutrino detectors that measure the number muon neutrinos coming from the atmosphere—as opposed to the ones traveling through the earth—show a smaller number traveling through the earth. [33 and 36] Why would this be? The Standard Model claims this is evidence of neutrino oscillation and therefore neutrinos must contain some mass. In the WPTL, this is evidence that neutrinos are nothing more than neutral radiation. Here is why. If one shines a light through a glass sphere, some of the light is absorbed by the glass, and a little less light travels through the sphere. The exact same thing is happening with the neutral neutrino radiation that travels through the earth; some is absorbed causing radioactive decay in the earth's interior. This leaves a slightly weaker background of neutrino radiation arriving to the detectors after traveling through the earth. This weaker field shows up as fewer muon neutrino induced decays. [35, 36 and 37]

The muon decay experiment clearly shows the direction from which more of the neutral neutrino radiation is coming from. [36] Remember momentum is conserved. This experiment also shows us that neutrino radiation in not as penetrating as now believed. Approximately ten parts in a million of muon neutrino radiation is absorbed traveling through the earth. The muon-neutrino decay experiments indicate this. If this is true, far fewer neutrinos would be detected coming from the sun because a significant percentage would be absorbed while traveling through the sun.

This background of neutral neutrino radiation that travels through the earth may have had an effect on the 1956 experiment showing a violation of parity in the decay of neutrons. [31, 32, 33 and 34] In this 1956 neutron decay experiment, slightly more electrons are ejected with a downward spin compared to the upward direction. Why should this be? Was parity violated? The reason more electrons may be ejected with downward spin could be because the cobalt atom's nuclei were held in place by a magnetic field. {71} If neutrons are a composite particle—the electron's spin around the proton and/or the neutron's spin would favor one spin over the other, when held in place by a magnetic field during the decay process. The favored downward spin could also be caused by the slightly more intense natural background of neutrino radiation in the downward direction as opposed to the upward direction. As in the muon neutrino decay experiment, some of the neutral neutrino radiation would be absorbed traveling through the earth when conducting these types of experiments.

In the WPTL, the different types of neutrinos (electron, muon, and tau) can be compared to different types of EMR like x-rays, radio waves, and gamma rays. The different types of neutral neutrino radiation are all composed of the same type of basic particles. Only the frequency and/or concentration of the waves of particles vary.

In the Standard Model, the radioactive decay process always ends in a loss of mass. But in some cases of radioactive decay, there is a gain of mass. Chlorine 37 to argon 37 is an example. In the radioactive decay processes that end with a gain of mass, it is claimed are the result of neutrino induced decays. Why would only radioactive decays that result in a gain of mass be neutrino induced? The odds would be even greater that most decays that result in a loss of mass would also be neutrino induced, because even less neutrino radiation (less energetic neutrinos) would be required to induce these types of decay. By viewing neutrino radiation as the source of most radioactive decays, one can easily explain what has happened to the missing neutrinos from the sun. A significant amount of the sun's neutrino radiation is absorbed inducing radioactive decay in the sun and some traveling through the earth before it ever reaches our neutrino detectors.

This—and the fact that neutrinos are radiation—account for the missing neutrinos we detect coming from the sun. Partial absorption of the neutrino radiation by neutrons in the nucleus accounts for some of the missing neutrinos. The fact that neutrino radiation is not as penetrating as current theory states also leaves us with less of a background of neutrino radiation.

Quantum mechanics does not have a good explanation for the energies of the decay products in the radioactive decay fission process. [39] There is no exact calculation possible as to why the decay products have these wide ranges of energies. If the decay resulted from neutrino radiation, it would explain this perfectly. The decay energies of the fission products cannot be calculated before the fission process takes place, because the exact energy of the neutrino radiation striking a neutron in the nucleus is unknown until after the collision. Quantum Mechanics also offers no explanation as to how a fundamental particle, the neutrino, can contain many different quantities of energy at the same velocity. In the WPTL the energy of the neutrino (radiation) can be in any quantity of Planck's constant in value.

But what about decays by way of alpha particles in which there is insufficient energy for the alpha particle to escape the nucleus? What about the quantum tunneling of the alpha particles in this decay process? This will be explained later. Next, let's discuss the strong nuclear force.

The Strong Nuclear Force

The strong nuclear force is an electromagnetic force. This will be contested, and rightly so by the scientific establishment. But in the WPTL, particles with mass have two parts, a particle with mass at the center and an electromagnetic field surrounding the particle. With this view of a proton it is quite easy to visualize how a nucleus with like charged and neutral particles can hold together.

Let's take a look at a simple element, helium 3. Helium 3 has the following arrangement of particles in its nucleus: two protons separated by a neutron. It can be represented like this:))P((()()EP()()))P((. The neutron ()()EP()() separates the two positively charged fields of the protons))P((. This reduces the repulsive force between the two protons. You could say the protons hide from each other's positive repulsive field in the shadow of the neutron. There is still a positive electromagnetic field surrounding the three nucleons. In fact, the positive field surrounding a nucleus increases with each proton added until the element iron. From then on, the positive field surrounding a nucleus weakens. One could view this as combining two eggs with the whites of the eggs surrounding the two combined yolks. This is where the neutrons come into play. They keep the yolks (protons) from directly feeling each other's repulsive positive charges.

This view explains why helium 4 has such a stable nucleus. There are two neutrons separating the two protons, and at the same time, the two protons can be in closer proximity, at a lower energy level, than they are in helium 3. Since the two protons are closer together, their positive-charged fields overlap more, which increases the strength of the positively charged field surrounding the nucleus. The helium 4 nucleus is shaped with the two neutrons at the two corners separating the protons completely. It can be represented like this:

))P((( )( )EP( )( )
()()EP()()))P((

Notice the two protons are now closer together but still shielded from each other's positive fields. The shape of a nucleus becomes more complicated in elements with more protons and neutrons. There are other factors that account for the shape of a nucleus; the spins of the particles for example. But with this new view of the strong nuclear force, one should be able to find the relative position of all the protons and neutrons in stable nuclei.

Another piece of evidence supporting the neutron shielding of the protons from each other's positive coulomb repulsive force is the fact that there is a steady need for more neutrons as the number of protons increase in a nucleus. It goes from roughly a one-to-one ratio to more than two-to-one in nuclei with a lot of protons. As the number of protons increase, the two-dimensional separation of the protons becomes a three-dimensional separation problem requiring additional neutrons.

This electromagnetic force is the same force that holds electrons in place (orbits) in atoms. In an atom, it is the positively charged nucleus that creates the negatively charged vacuum or neutrally charged field toward the center of an atom. The electrons in an atom are not attracted to the positively charged nucleus; they are repelled by their own unbalanced negatively charged fields. The negatively charged fields are greater on the side of the electrons facing away from the nucleus. The side facing the nucleus is neutralized somewhat by the positive electromagnetic field created by the nucleus. This is why the electrons are forced into the lowest possible quantum energy level by the more negative field on the other side of the electrons.

Now wait a minute. Why is the strong nuclear force so much stronger than the electromagnetic force? It isn't any stronger. With the strong nuclear force, we are measuring the electromagnetic force in an extremely short distance when compared to the size of the atom. The atom is about a million times the size of its nucleus. It is not unreasonable to assume the positive electromagnetic force surrounding the nucleus would then be a million times more intense than the negative electromagnetic force created by the electrons holding them in place. In fact, the difference between the strong force and the electromagnetic force is about a million to one. Protons in the nucleus create an intensely concentrated positive field surrounding it. This is why it has so much strength. This strong positive charge is the strong nuclear force.

For example, the size of an atomic nucleus is equal to 10^{-14} meters. The diameter of an atom is about 10^{-8} meters. [41] The energy required to ionize a hydrogen atom is 13.6 electron volts (eV). [40] From the size of an atom and from the energy required to ionize the hydrogen atom we can arrive at the coulomb repulsive force between two protons at $.5 \times 10^{-8}$ meters, the radius of an atom. The coulomb repulsive between two protons at $.5 \times 10^{-8}$ meters would equal the ionizing energy of the hydrogen atom or 13.6 eV. This is a very tiny amount, but this is also at a very large distance compared to the size of a nucleus. Reducing the distance by a factor of one thousand to $.5 \times 10^{-11}$ meters would increase the coulomb repulsive force by a factor of one million (it increases by the square of one thousand), and the coulomb repulsive charge increases to 13.6 million electron volts. There is still a large distance between the two protons, in fact about 200 times the width of an atomic nucleus. It is plain to see if one reduced the distance between the two protons by another factor of ten, giving a new distance between them of $.5 \times 10^{-12}$ meters, the 13.6 million electron volts would increase by the square of ten, or one hundred, and now equal 1.36 billion electron volts. This is still at a distance of about 20 times the diameter of a nucleus.

The nuclear binding energy (strong force) of the helium 4 nucleus is 28.3 million electron volts. [45] This is down right paltry compared to the coulomb repulsive force between two protons at a distance of 20 times the diameter of this helium nucleus.

There is another strange thing about the strong force holding the nucleus of an atom together. Why does the strong force turn into a strong repulsive force when the nucleons are forced too closely together? The Standard Model cannot easily explain this. This is more proof that the neutrons are in fact separating and shielding the protons from each other's positive coulomb repulsive charges. When they are forced to closely together, the neutrons doing the shielding are forced out from between the protons exposing the strong repulsive force between the protons. So the strong attractive nuclear force turns into a strong repulsive force over a distance of about one femtometer, about the diameter of a neutron. [46 and 48]

If this view of the strong nuclear force is correct, it creates one very big difference in the fusion process. For all practical purposes, it would be impossible to fuse two protons; the union would be unstable at anything much above absolute zero. In the Standard Model the sun is not hot enough to fuse two protons, so quantum tunneling of protons is required. [47, 42 and 43] In the WPTL a neutron must first be created from an electron and proton. This neutron then fuses with a proton in a hydrogen atom creating a deuterium atom (it can also fuse with other nuclei). Two deuterium atoms then fuse creating the helium atom. In fact, according to this theory that is what prevents stars like our sun from exploding, the difficult and random process of fusing an electron and proton. Above a certain pressure, electrons and protons collapse into neutrons in a cascade, creating a supernova and its remnants, a neutron star.

But this creates another problem. If the strong nuclear force is just an electromagnetic force, what binds the neutrons to the nucleus? They are only weakly bound by a weak magnetic force to a nucleus. There are two other reasons they stay in a nucleus: their low kinetic energy level and the fact that their positions are mostly in the interior of the nucleus between protons. You could say they are content to stay where they are. To free them, energy must be added to overcome their inertia and weak magnetic attachment to the nucleus. In fact, when neutrons fuse with a nucleus it is not the lack of energy that gets them to fuse, it is the fact they possess too much kinetic energy to fuse. They bounce away losing some of their kinetic energy with each collision and finally fuse when most of their kinetic energy is gone. [69 and 70]

If the strong nuclear force acts equally on protons and neutrons, why is it so much easier for neutrons to escape from the nucleus? The ease with which radioactive atoms decay by neutron emission proves it is much easier for a neutron to escape from the nucleus. The chain reaction in an atomic bomb also proves this.

This creates another question. Where does the mass loss in the fusion process come from? Earlier, it was stated that mass comes from the never-ending process of basic particles exchanging angular momentum. There is a second type of mass, which could be referred to as gyro mass. This second type of mass is derived from the same origin as the exchanging of angular momentum origin of mass. Linear momentum in the form of angular momentum is the origin of mass. Angular momentum is also the origin of a gyroscope's inertia or resistance to changes in direction.

The electromagnetic fields surrounding particles with mass are composed of basic particles that have converted most of their linear momentum into angular momentum. These rotating basic particles resist moving for anything that comes into contact with them. They appear to have mass. It is this form of mass that is radiated away when electrons or protons are forced by their own electromagnetic fields into a lower energy level. They actually radiate away some of their own electromagnetic field. This effect is barely noticeable when an atom captures and electron and radiates away EMR, but when a nucleus captures an electron, proton, or neutron, the extremely concentrated electromagnetic field surrounding the nucleus forces the newly captured particle into a much lower energy level. The electromagnetic field surrounding this newly acquired particle has a significant amount of its electromagnetic field radiated away. The total mass of an electron, proton, or neutron is the sum of the mass of the particle and the electromagnetic field surrounding it.

There is another thing worth noting: if the strong force is just an electromagnetic force that has only a small magnetic effect on a neutron, why would a nucleus that captures a neutron release energy and drop to a lower energy level? A captured neutron permits two protons to drop to a lower energy level because this captured neutron adds neutral shielding between these two protons. This permits these two protons to be forced into a lower energy level by the positive coulomb force surrounding this nucleus.

Quantum Tunneling of Alpha Particles

Let's return to quantum tunneling of alpha particles in certain decay processes. Why is quantum tunneling necessary? [42 and 43] It is necessary because in many decay processes the alpha particle lacks the kinetic energy to penetrate the positive coulomb barrier surrounding the nucleus. The very fact that this type of quantum tunneling is necessary proves the strong nuclear force is a positive electromagnetic force. If the coulomb barrier can keep the alpha particle from escaping, why do you need a strong nuclear force holding the nucleus together? You don't. The strong nuclear force and the coulomb barrier are one and the same.

Then how does the alpha particle escape during this type of decay process? It does not need to tunnel; it goes straight through by classical means. Assume neutral neutrino radiation of sufficient energy strikes an alpha particle-like arrangement in uranium. This alpha particle bounces off the rest of the nucleus with less kinetic energy than needed to penetrate the coulomb barrier surrounding the uranium nucleus. One would assume it could not escape, but it does. How? It can escape because the strong nuclear force is an electromagnetic force. As the alpha particle bounces off the nucleus, the positive coulombs repulsive force between it and the nucleus increases. At the same time, this alpha particle is penetrating the coulomb barrier, surrounding the nucleus, its repulsive force diminishes. As the alpha particle penetrates the coulomb barrier, the part it has already passed through, actually starts pushing it out. For this reason, the alpha particle can escape with less kinetic energy than is needed to penetrate the coulomb barrier going from the outside into the nucleus.

Wouldn't the same process happen with an alpha particle entering the nucleus? It does, but there is a difference. When this alpha particle exits the nucleus, it pushes the surrounding positive coulomb barrier out into a larger area which weakens the coulomb barrier slightly. Just the opposite happens when an alpha particle tries to enter into a nucleus. The coulomb barrier is forced into a smaller area concentrating it. For this reason and the previously mentioned reason, the alpha particle can exit a nucleus with less than half the kinetic energy it needs to enter into this same nucleus. [44] The coulomb barrier is greater entering than exiting a nucleus. Quantum tunneling proves this.

The Atom

With these different views of particles and forces, let's take a new look at the atom. Take the simplest atom, hydrogen. Why doesn't the electron fall into the proton? They have opposite charges, and one would think this would force them together. The reason this doesn't happen is because the proton and electron contain heat. This heat is stored potential energy in the form of the electron and proton oscillating around inside their electromagnetic fields. The frequency of the electron's and proton's oscillations must also be in a harmonizing frequency with each other in the hydrogen atom. This is true for all atoms. The electrons and protons at different energy levels must oscillate in harmonizing frequencies with each other. Only electrons in atoms, and protons in the nucleus, at the same energy level, can oscillate at the same frequency.

The proton is 1836 times as massive as the electron, and the oscillation frequency in the form of heat it contains must be in a harmonizing multiple of what the electron contains. This only permits certain harmonizing frequencies. This is the origin of permissible frequencies atoms radiates in the form of EMR. Energy levels between two frequencies in harmony are radiated away, and the electron drops to a lower stable energy level, which is in resonance with the proton's oscillation. Since the mass of the electron is so much smaller than the proton's, and the electron's velocity is much greater than the proton's, the electron is the one that radiates away the excess energy and drops to a lower energy level.

It is this disparity in the oscillating velocity and frequency between the electron and proton that keeps them from falling into each other. The electron with its much higher oscillating velocity dances all around the proton in the hydrogen atom. In atoms with more than one proton, the two inner electrons keep each other from falling into the nucleus. At the lowest permissible energy level, these two inner electrons are oscillating at the same frequency and are resting in a stable balanced energy level. In atoms with two or more electrons, the electrons pair up in a stable frequency by energy levels or orbits.

A similar thing happens in the nucleus. It becomes more complicated because the nucleus contains protons and neutrons, which have a slightly different mass and a different charge. The protons and neutrons pair up in the nucleus, and the shell or orbital energy levels of electrons in an atom can describe the nucleus fairly accurately. [49 and 67] And why shouldn't it? Both originate from the electromagnetic force.

It is the harmonization of electrons, protons, and neutrons forming pairs that creates the difference between boson and fermion atoms.

One more thing: under certain conditions a proton and electron can and do fuse, creating a neutron. Now, we don't understand all the factors in this process yet, but the secret to virtually limitless nuclear energy could come from solving this problem.

One naturally tends to think that the higher the oscillating frequency of a particle, the higher the energy level it is in or the more potential kinetic energy this particle contains. This would be true under equal pressures between particles with equal mass. One easy way to see how a high oscillating frequency can be a lower energy level is to drop a rubber ball from six feet. We know the ball contains more potential kinetic energy at the dropping point of six feet then it will in a return bounce of three feet or one foot. So the kinetic energy contained by the bouncing ball is reduced with each lower bounce. We also can hear with our own ears the frequency of the bouncing increase as the ball bounces lower and lower until it stops bouncing. So in the case of the bouncing ball, the kinetic energy is dropping at the same time the frequency of the bounces increases. This is why things tend to get a little complicated in atoms. With electrons and protons, we not only have to consider the particle's oscillating velocity and mass, but we must take into account the pressure (binding or ionizing energy) containing this particle to get a true picture of the potential kinetic energy stored.

Inner electrons in atoms are at a lower energy level. That is they contain less potential kinetic energy in the form of heat, but they oscillate at a higher frequency. This is where pressure comes into play. The electromagnetic force holding electrons in place is stronger nearer the positive nucleus. So the electromagnetic charge between the electron and the nucleus increases the pressure the electron is under forcing it into a higher frequency. This higher frequency may not be in harmony with the rest of the atom, so the extra energy must either be passed to other parts of the atom or radiated away in the form of x-rays. [68] Now these inner electrons would have a higher oscillating frequency because they are bound by a stronger electromagnetic field, but they are still at a lower energy level. This is similar to a rubber ball attached to a paddle. If you were to bounce the ball off the wall and move the paddle toward the wall at the same time, the frequency of the bouncing ball increases at the same time as the potential energy is being "squeezed" from the ball.

The point being made here is that temperature and heat, or potential kinetic energy, are not the same. This explains why inner electrons, which are at lower potential energy levels in heavier elements, radiate away more energetic forms of EMR while electrons in the lighter gases at a much higher energy level radiate away longer, lower energy frequencies of EMR even though they contain more potential kinetic energy in the form of heat. The WPTL also maintains that the superwaves of lower frequencies radiate away more individual waves (1/2h particles) than higher frequencies per oscillation.

The nucleons inside the nucleus of a helium atom contain about a million times less potential kinetic energy in the form of heat compared to the proton in a hydrogen atom. Why is this? The electromagnetic pressure holding electrons in place in atoms is about a million times weaker than the electromagnetic pressure holding protons inside the nucleus. One could view this as similar to the way a compressor forces heat out of a gas as it is compressed into a liquid. The nucleons inside the nucleus could be viewed at a high temperature, that is high oscillating frequency, but at the same time containing very little heat. In fact, if the positive electromagnetic nuclear pressure holding the protons in a nucleus could magically disappear, the protons and neutrons in the iron atom's nucleus would be close to absolute zero. The nucleons in an iron atom's ground state are at the lowest energy level of any atom. They are also confined at the highest electromagnetic pressure for nucleons in any atom.

This increase in pressure is why the so-called strong nuclear force radiates away about one million times more energy than the chemical electromagnetic force holding electrons in their atomic places. This energy must be returned to free protons from a nucleus. Enough energy must be added to an individual proton to overcome about half the positive nuclear coulomb barrier in order to free it, and enough energy must be added to a neutron to break the weak magnetic force and its inertia holding it to the nucleus. But a proton must contain enough kinetic energy to overcome the positive coulomb barrier to free another proton from a nucleus. This leads to a strange point: a proton must also contain the same amount of kinetic energy to free a neutron from the nucleus of an atom. The reason is that the proton must overcome the positive coulomb barrier before it can even collide with the proton or neutron in the nucleus. This creates a false appearance that the strong nuclear force is acting on the proton and neutron equally. It is not. The strong nuclear force—the positive electromagnetic force—actually has only a small magnetic effect on the neutron.

Now a neutron can free another neutron at a much lower kinetic energy level, but it would have to contain about half the kinetic energy of the coulomb barrier to free a proton from the nucleus. A positively charged particle needs about half the kinetic energy to escape a nucleus compared to what is needed to enter into the same nucleus. This is because of the quantum tunneling effect explained earlier.

Molecules

Molecules are two or more atoms joined by the electromagnetic force. One may wonder how atoms form electrical bonds when the two or more atoms forming these bonds are electrically neutral. Atoms, unless ionized, are electrically neutral. But there are negative- and positive-charged spots on the outside surface of atoms. It is these oppositely charged spots that form the chemical bonds between atoms. Atoms are rarely found individually in nature at low temperatures. They form compounds with other elements or among themselves. This is a lower energy state, and nature constantly radiates away energy and drops to a lower energy state. Molecules are at a lower energy state than atoms because they are more balanced electrically. That is they have fewer and smaller areas of positively or negatively charged spots on their outer surfaces.

Another way of viewing atoms and molecules is to view them as having shape. It is generally accepted that molecules have a shape. [50] In the WPTL, atoms also have shapes; they are not perfectly spherical. As mentioned earlier, they have negative and positive spots on their surface. Also mentioned in a previous chapter, viewing electrons in orbit around atoms is misleading; the electrons do not actually orbit in an atom. [55] They are in a relatively fixed position in relationship to each other. This is what gives atoms their shape, which in turn gives shape to molecules. The chemical characteristics of atoms originate from the shape of the electromagnetic field surrounding them, which is shaped by the positions of the outer electrons. If atoms do not have shapes, where would the shape of molecules come from?

Gravity

We need to take another look at the basic particles everything in the universe is made of. When these basic particles collide, they exchange linear momentum if they have opposite spins. They must exchange linear and angular momentum if they collide with the same spins. It takes one extra quanta second of time to exchange linear and angular momentum. In the one quanta second that the angular momentum is exchanged, one quanta of mass and time are created. This time is directly involved in the creation of the force of gravity.

Particles with mass—protons, neutrons, and electrons—are bombarded by swarms of basic particles from all directions. These basic particles collect around particles with mass because of collisions that exchange spin. Collisions that exchange spin take time and this increases the number of basic particles, where particles with mass happen to be. The more mass, the more collisions, because for every collision, the number of basic particles is increased by one for a quanta second in this area. The number of basic particles above the normal background of basic particles is directly related to the amount of mass in any given area and would drop off from the center of any mass by the inverse square rule. A gravity field also weakens by the inverse square rule in the distance from the center of this mass. (This is close enough, but not exactly correct, more later.)

It takes energy to move the colliding basic particles in the moment they are exchanging angular momentum with the basic particles in particles with mass. These collisions that exchange spin turn dark linear energy in these basic particles into dark mass for one quanta second.

These collisions also create time as we know it. The greater the collection of particles with mass, the greater the number of collisions between these basic particles. It is this created time and heat that gives us the force of gravity. Yes, heat. A particle with mass must contain a minute trace of heat to move toward the center of a gravity field. Any amount of heat at all will do, but at absolute zero, an object with mass would become weightless. It would still contain the same mass as before, but would no longer feel the relentless tug of gravity.

Any motion created by gravity comes from the heat contained by particles with mass. But gravity does not affect the temperature of mass in any way. The very physical space-time shape of particles with mass are changed. Time runs slightly slower on the side of particles with mass that face the center of a gravity field. It is as Einstein said: a gravity field changes the very shape of space-time. [62] The actual geometry of space is unchanged; only the time in space-time is changed. Time runs slower toward the center of a gravity field.

Since time runs slower toward the center of a gravity field, particles with mass that contain even trace amounts of heat spend a trace amount of increased time, from their heat-induced oscillation, toward the center of the gravity field. This causes the particles with mass to fall toward the center of a gravity field when free floating. It is very similar to the way a Mexican jumping bean is able to turn an internal oscillation into a linear movement. The difference is that the internal oscillation or heat in not used up or turned into linear movement (kinetic energy) in a gravity field. That energy comes directly from the dark energy field of the swarming basic particles. Earlier it was mentioned that when two basic particles with the same spin collide, they are one quanta second of space-time closer together. It was also mentioned that this collision turned kinetic energy into one quanta of mass for one quanta second. To move out away from a gravity field, this energy turned into mass must be replaced. One might wonder how this could be when talking about atomic sizes of particles with mass. The difference is minute (incredibly small) from one side of a particle such as an electron to the other, but the quantity of basic particles that create the force of gravity are enormous. They interact, depending on the strength of the gravity field, with the approximately 10^{22} of basic particles in each electron.

So the energy gained by an apple falling to the ground comes from dark energy. To place the apple back into the tree, you must replace the dark energy it gained in the form of kinetic energy. That's work. Neither Newton's nor Einstein's theory of gravity can explain where the energy gained or lost in a gravity field comes from. Neither can they explain why we mortals feel the constant pressure of a gravity field. We feel the constant pressure because every particle with mass and heat in our body spends a little more time toward the center of the earth than it does away from the center.

A natural thing to question is that since gravity needs heat in the form of oscillating particles with mass to be attracted toward the center of a gravity field, why does the acceleration due to gravity affect all particles with mass equally regardless as to what their temperature is? It is not the temperature that causes mass to be attracted toward each other. It is the increase in time these particles spend toward the center of the gravity field. Increase their temperature and you increase the frequency of the oscillations, which simply divides this increase in time into smaller segments. The total time increase is the same regardless of the temperature.

There is an experiment that may be able to confirm this theory of gravity. By cooling fermion atoms extremely close to absolute zero, the weight of these fermion atoms may become a little less and fall a little slower. The reason; since fermions cannot enter the same quantum state, some of the atoms (or parts) may enter the absolute zero state for fleeting quanta seconds.

There is another experiment that may be able to prove this theory of gravity. Since gravity is the exchange of basic particles (gravitons) at the velocity of light, the moon should move slightly away from the earth at the beginning of a total eclipse of the moon. The reason is because the sun's gravity field has to travel through the earth during these eclipses. Since the earth is a dense mass containing a field of many times the number of basic particles compared the empty space on either side, the earth would slow down the number of gravitons reaching the moon that originated and creates the sun's gravity field. Once the mid point of the eclipse is reached, just the opposite occurs; the moon is drawn back toward the earth. The moon slowly returns close to its original orbit as the eclipse fades. It ends up in nearly, but not the exact orbit it had before the eclipse. It should recede from the earth by a very slight amount because of the time it takes for the gravitons to travel through the earth and reach the moon. About one second's worth of the gravitons from the sun's gravity field would be carried off to the side of the moon into the rest of our galaxy. Some of the sun's gravity would be lost to the moon forever during each eclipse.

Let us take this new view of gravity and see how it might explain some of the things observed in the universe. For example, how much would you weight at the center of the earth? Your weight would be zero. The gravitational pressure (time) would be equal on all sides of you. If the earth were hollow, you could float around weightlessly inside. But you would be slightly more massive; that is you would have more inertia. The reason is that the basic particles from which the force of gravity arises would be at their densest at the center of the earth. Therefore, more collisions between the basic particles turns more linear momentum into mass. This would increase the inertia your body contains by a very small amount. Time would also run slightly slower than it would on the surface of the earth.

It is generally accepted that the velocity of light travels slightly slower in strong gravity fields. Why? Because EMR is a wave of the same basic particles that create gravity, and in strong gravity fields there are more of these basic particles. Therefore, more of the time the radiation would stop its forward motion to exchange this angular momentum with gravitons.

It would also slightly change the strength of gravity at extremely long distances. The force of gravity would no longer be infinite. All traces of gravity, or the basic particles that create gravity, would eventually come under the influence or control of more local masses. Gravity would actually decrease by slightly more than the inverse square rule over great distances (billions of light-years). But a gravity field would be slightly stronger at the center of our solar system and in the center of galaxies than the present theory allows for. Why? It has to do with dark energy turned into dark mass. At the center of galaxies, there is a significant amount of unseen dark mass created by these basic particles in the process of colliding and exchanging spin. At the moment the spin is exchanged, one quanta of mass is created for one quanta second. This is where the dark mass comes from.

After having said the gravity field is stronger in the center of the earth and a galaxy, the force of a gravity field is weaker in the center of a galaxy and our solar system, just as it is in the center of the earth. The force of gravity is zero in the center of the earth, the center of the sun, and the center of a galaxy. While the force of gravity would be zero at the center of the field, the gravity field would be at its most concentrated. That is, its effect on weight would be zero, but its effect on time and mass would be at its maximum.

Now we can use this new view of gravity to explain why stars rotate faster at the outer reaches of a galaxy than the present theory would allow for based on the mass of this galaxy. The outer stars in a spiral galaxy do not rotate faster than the mass of the galaxy allows for; it is the inner stars that are rotating too slowly. As mentioned earlier, the force of a gravity field weakens as one nears the center because the mass on the outside is now exerting a force pulling away from the center of this gravity field. So the strength of a galaxy's gravity field is subtracted from the stars nearer the center by the stars farther out until the force of gravity reaches zero at the very center of a galaxy. This causes the stars toward the center of a galaxy to rotate much slower than they would rotate without stars farther from the center of a galaxy counter balancing the force of gravity from the stars on the other side of this galaxy. This very same effect would happen here on earth. The orbital velocity of a satellite would actually decrease below the surface of the earth until it would reach zero at the very center. The preceding explanation could also explain why some galaxies and star clusters are apparently stable with most stars having the same rotation velocity, or apparently none, for extended periods of time.

In summary, there are three additional factors, other than mass alone, affecting the strength of a gravity field: (1) a gravity field drops off slightly faster than the inverse square rule over extremely great distances because of single graviton use; (2) a gravity field (not the force) is slightly stronger because of dark unseen mass, especially toward the center; (3) all mass farther out from the center of a gravity field must be subtracted from the force of this field at any given point of measurement.

Could this view of gravity explain the anomalous acceleration in the direction of the sun noticed in the Pioneer spacecraft? [51] The force of gravity is subtracted by mass farther from the center of a field at any given point of measurement. As a moving spacecraft leaves our solar system, the force of the sun's gravity is reduced by the inverse square rule, but it increases by double the amount of mass passed by on the spacecraft's exit from our solar system. The gravitational pull of the asteroid belt and larger outer planets turns into an additional gravitational drag as the Pioneer spacecraft passes beyond these bodies of mass. It is not an equal exchange. A subtraction becomes a plus; the gain is double the gravitational force from the asteroid belt and the outer planets as the spacecraft passes them by on their exit from the solar system.

This theory of gravity does not rule out black holes, but it does suggest they would not exist at the center of galaxies, because the force of the galaxy's gravity would be zero in its center. It is believed in the WPTL that at the very center of a galaxy, an extremely strong electromagnetic field rules this nearly weightless environment. It could actually be strong enough to rip stars apart and expel the charged ions out of the poles of the galaxy at extremely high velocities. This could be the way galaxies recycle themselves. A galaxy's electromagnetic field would transmit these charged ions at extremely high velocities to their outer edges. Could extremely high energy comic rays be evidence of this?

4
Conclusions

General Relativity and Quantum Mechanics

The Waves of Particles Theory of Light (WPTL) does not make Quantum Mechanics or The Theory of Relativity false or obsolete. It adds to both of those theories while placing some limits or constraints on them. It also brings both theories into an understanding with classical physics. The mathematics of quantum mechanics and relativity has been very effective in our understanding of the basic micro- and macro-worlds.

But there are some definite differences between these theories. For one, there isn't any quantum entanglement beyond gravity and/or the electromagnetic force in the WPTL. Secondly, there is no superposition. A single electron oscillated at an extremely high frequency between two points creates a bone-shaped electromagnetic field around itself. It only appears to be in two places at the same time, but it never is. A third difference is in the way electromagnetic radiation (EMR) is radiated and absorbed. Quantum mechanics views a single wave of EMR as expanding out into infinity, and when it is absorbed, the whole wave collapses. In the WPTL, a single wave has a cross section of a Planck length squared. Even a superwave, which contains approximately one billion single waves, is not usually absorbed in whole units. Radio superwaves are never absorbed in this manner.

There is also a difference in frames of reference. In Einstein's theory of relativity, there are no preferred frames of reference. In the WPTL, it would appear there are preferred frames of reference. [66] Both theories agree that the speed of light is absolute and cannot be added to or subtracted from. If that is true, then wouldn't this statement be true: there is an absolute frame of reference in this universe, and it is the actual velocity of light. The irony is that even that statement is not the complete truth, the velocity of light is also dependent on the strength of the gravity field it is traveling through. An absolute frame of reference like absolute zero may be unattainable, but like with absolute zero, we can get very close. The cosmic background radiation lends further proof that there is a preferred frame of reference. [23] It could also be proven by measuring the apparent one-way velocity of light in both directions from the earth. [65] The Global Positioning System (GPS) must take these apparent velocities into account in fixing positions. This is known as the Sagnac Effect.[52]

If simultaneity could not be established, how could the GPS be able to find exact locations? If simultaneity is established, a preferred frame of reference is also established. Now one can imagine time frames in which one could not establish simultaneity or a preferred frame of reference, but they must be out of this universe. Nowhere in this universe can two spaceships pass each other without knowing which is moving, which is stationary, or even their own velocity and the other spaceship's velocity. How can this be determined? One way would be to measure your own spaceship's velocity against the cosmic background radiation. [23]

The earth is moving relative to the cosmic background radiation at some 370 miles per second. [23] This also does not support the Big Bang theory. This is a relatively slow velocity, only a little more then the velocity of our solar system around the Milky Way galaxy.

There is other evidence that there are preferred frames of reference and quite possibly an absolute frame of reference. Imagine launching satellites from an earth-only universe. It would not be long before we realized there was a preferred direction in which to launch these satellites, and we would soon learn how fast the earth was rotating. This same effect is used to launch satellites to the outer planets, and it would certainly be used to explore the galaxy.

It is not entirely correct to say the Michelson-Morley experiment proved there was no ether. [63] While the WPTL maintains there is no need for ether, it does maintain that there is an apparent difference in light velocities measured from a moving object, depending on the direction of measurement. [65] The Michelson-Morley experiment measured the round-trip velocities of light. Round-trip velocities of light will not show a difference even if there is an apparent measurable difference depending on the direction of travel. This seems like a contradiction, but it is not if one really believes the velocity of light is absolute, which it is. The WPTL has a simple explanation as to why one cannot add to or subtract from the velocity of light: each half-wave of EMR is radiated and absorbed in the smallest unit of time possible, one Planck unit of time. That is the reason the motion of the radiating body cannot add to or subtract from the velocity of light. And this is why light is red-shifted and blue-shifted when radiated from a moving source.

The WPTL can explain in a classical physics way why time is dilated and why a particle gains mass as it approaches the velocity of light. As an electron or proton is accelerated, it collides with basic particles in EMR, electromagnetic fields and gravitons from a gravity field. Some of these basic particles that collide in the same time frame as the electron or proton become part of them. At lower velocities, nearly all of these added basic particles are radiated away. But as the electron or proton approach the velocity of light, there is less time to radiate away these basic particles until at the velocity of light there is no time left to radiate away these basic particles. This effect happens in a parabolic curve because as the velocity increases, collisions increase while the time to radiate away these basic particles decreases. So the electron or proton increases in mass and momentum. At the velocity of light, the electron or proton cannot be accelerated any faster, because the electromagnetic fields used to doing the accelerating do not work faster than the speed of light. But if one keeps trying to increase the velocity of the electron or proton, they only collect more and more basic particles, which only increases their mass and momentum. There is not theoretical upper limit on how massive a single electron or proton could become.

In the WPTL, all the other particles created in particle accelerators are formed from the excess mass of high velocity electrons, protons, and neutrons contained during collisions. Why certain sized particles? It probably has something to do with certain sizes that resonate internally in a semi-harmonizing fashion. But none of these created particles are stable. They all decay in short order to electrons, protons, or their antiparticles, as well as into EMR and neutral neutrino radiation. In the WPTL, none of these created particles have anything to do with the weak or strong nuclear force. With powerful enough particle accelerators, the Higgs boson will be found. [64] It has nothing to do with mass or gravity in this theory.

The subatomic particle zoo can get extremely complicated, because in addition to the mass gains, there are electromagnetic field complications. For example, could a muon be an electron with it own, as well as a super sized neutron's, neutral fields surrounding it? Maybe this is why it is so penetrating even with a charge. If so, this would also explain why it radiates away neutral neutrino radiation as it decays.

The WPTL also has no need for quarks or gluons. It simplifies the Standard Model by a thousand-fold. The Standard Model and Quantum Mechanics are beautiful mathematical models. The ancient Greeks also had a beautiful mathematical model to predict where the moon and planets would be at some future date. [53] It worked quite well, but it was totally devoid of reality. As a theory, it had no basis; it was just a mathematical model. One might ask where is the math in this theory. Where is the math in the theory of evolution? The math in the WPTL is as simple as zero plus one—on or off.

Can The Waves of Particles Theory of Light ever be proven to be correct? Maybe, if some of the more unusual tests mentioned prove to be correct, it could lend support to the theory. That would definitely be a plus, but the ultimate proof may be that it is the simplest way to explain what we observe in the universe. Remember, when nature is confronted with two courses of action, it always takes the easy road, the path of least resistance. Can anything explain the particles and forces we observe in nature simpler than whether two basic particles exchange angular momentum or not?

References

[1] What is the Photoelectric Effect? PhysLink.com
http://www.physlink.com/Education/AskExperts/ae24.cfm?CFID=16882158&CFTOKEN=93301688

[2] Abstract: Laser: The photoelectric effect is not related to the frequency of incident light.
"http://adsabs.harvard.edu/abs/1989STIN … 8923869G"

[3] Anomalous Multiphoton Photoelectric Effect in Ultrashort Time Scales
http://scitation.aip.org/getabs/servlet/GetabsServlet?prog=normal&id=PRLTAO000095000014147401000001&idtype=cvips&gifs=yes

[4] Electron-positron pair production Information from Answers.com
http://www.answers.com/topic/electron-positron-pair-production

[5] The transmission of sub-wavelength hole arrays in **silver** films.
"http://adsabs.harvard.edu/abs/2005APS.SES.FB003W"

[6] Light transmission through sub-wavelength holes.
"http://www.nature.com/nature/journal/v391/n6668/full/391667a0.html"

[7] Beaming Light from a Subwavelength Aperture
"http://www.sciencemag.org/cgi/content/abstract/297/5582/820"

[8] Conservation of energy Definition and Much More from Answers.com
"http://www.answers.com/topic/conservation-of-energyPhysicsWeb"

[9] The reality of negative refraction.
"http://physicsweb.org/articles/world/16/5/3Red"

[10] light debut for exotic 'metamaterial'—tech—18 December 2006—New Scientist
"http://www.newscientisttech.com/article/dn10816.html"

[11] APOD 2000 May 26—Solar Sail
"http://apod.nasa.gov/apod/ap000526.html"

[12] Solar sailing 'breaks laws of physics'—04 July 2003—New Scientist
"http://www.newscientist.com/article.ns?id=dn3895"

[13] Compton effect Definition and Much More from Answers.com
"http://www.answers.com/topic/compton-effect"

[14] Bose–Einstein condensate Information from Answers.com
"http://www.answers.com/topic/bose-einstein-condensate-4"

[15] The Effects of Dispersion on High-speed Fiber Optic Data Transmission Fiber Ban
"http://www.fiber-optics.info/articles/dispersion.htm"

[16] The Big Bang Theory
"http://liftoff.msfc.nasa.gov/academy/universe/b_bang.html"

[17] The Steady-State Theory of the Universe
"http://www.schoolsobservatory.org.uk/study/sci/cosmo/internal/steady.htm"
"http://www.schoolsobservatory.org.uk/study/sci/cosmo/internal/steady.htm"

[18] Distant Supernovae
"http://apod.nasa.gov/apod/ap980114.html"

[19] Distant Supernovae Indicate ever-Expanding Universe ESO—Astronomers Contrivute towards Resolution of Cosmic Puzzle
"http://www.eso.org/public/outreach/press-rel/pr-1998/pr-21-98.html"
"http://www.eso.org/public/outreach/press-rel/pr-1998/pr-21-98.html"

[20] Supernovae Dimmer than Redshift would Suggest
"http://209.85.165.104/custom?q=cache:gI5SFUCVOaMJ:snap.lbl.gov/pubdocs/snap_synopsis.pdf+Distant+supernovae+brightness+puzzle&hl=en&ct=clnk&cd=13&gl=us&client=pub-5123197681197851"

[21] Absolute Brightness rises at first, then drops with distance.
"http://www.lbl.gov/Science-Articles/Archive/sabl/2005/October/04-supernovae.html"

References 65

[22] Cosmic Background Radiation
"http://hyperphysics.phy-astr.gsu.edu/hbase/bkg3k.html"

[23] ASTROPHYSICAL CONSTANTS—Local group velocity with respect to the CBR 627 kilometers per second.
"http://209.85.165.104/custom?q=cache:88t-Q9Hzm5IJ:pdg.lbl.gov/2002/astroonepagerpp.pdf+Earth+velocity+CBR&hl=en&ct=clnk&cd= 26&gl=us&client=pub-5123197681197851"

[24] BBC News SCI-TECH Universe 'proven flat'
"http://news.bbc.co.uk/1/hi/sci/tech/727073.stm"

[25] Hubble Measures the Expanding Universe
"http://science.nasa.gov/newhome/headlines/ast25may99_1.htm"

[26] The Accelerating Universe and the SuperNova Acceleration Probe
"http://209.85.165.104/custom?q=cache:QzuygR9IkF8J:snap.lbl.gov/ota/sfsutalk.ppt+expanding+universe+acceleration&hl=en&ct=clnk&cd= 8&gl=us&client=pub-5123197681197851"

[27] WMAP Cosmology 101 Inflationary Universe
"http://map.gsfc.nasa.gov/m_uni/uni_101inflation.html"

[28] A Guide to Entropy and the Second Law of Thermodynamics
"http://209.85.165.104/custom?q=cache:QteO_yrtd-8J:www.ams.org/notices/199805/lieb.pdf+Universal+entropy&hl=en&ct=clnk&cd=20&gl=us&client =pub-5123197681197851"

[29] Planck length 1.6160×10^{-35} meter
Planck time 5.3906×10^{-44} second
"http://www.planck.com/planckunits.htm"

[30] Van de Graaff Generators Construction & Demonstrations
"http://amasci.com/emotor/vdg.html"

[31] Parity Violation in Nuclear Physics—more electrons emitted in one direction than the other
"http://209.85.165.104/custom?q=cache:JsW5vCQ5o5cJ: www.fas.org/sgp/othergov/doe/lanl/pubs/00285652.pdf

+neutron+decay+percentage+violate+parity&hl=en&ct=clnk&cd
=3&gl=us&client=pub-5123197681197851"

[32] Neutron decay asymmetry—neutrons vertical spins—up and down decay asymmetry
"http://www.encyclopedia.com/doc/1G1-138950382.html"

[33] symmetry—May 2005—The Elusive Neutrino—a deficit was observed for atmospheric neutrinos originating on the opposite side of the Earth, traveling the largest distance.
"http://www.symmetrymagazine.org/cms/?pid=1000112"

[34] Guide to Nobel Prize—beta decay prefers the down direction for spin
"http://www.school.eb.com/nobelprize/article-48296"

[35] THE DISCOVERY OF NEUTRINO OSCILLATIONS—muon neutrino flux, with a larger deficit in the up coming direction.
"http://www.phys.hawaii.edu/~jgl/nuosc_story.html"

[36] American Scientist Online—Neutrino Oscillations—it is the upward-going muon neutrinos that are missing
"http://www.americanscientist.org/template/AssetDetail/assetid/15669/page/6;jsessionid=aaa5LVF0"

[37] South Pole Neutrino Detector Could Yield Evidences of String Theory—reducing the number of 'up' neutrinos?
http://physorg.com/news10295.html

[38] Solar neutrino problem—Wikipedia, the free encyclopedia
"http://en.wikipedia.org/wiki/Solar_neutrino_problem"

[39] Chapter 3 Radioactivity—energy can be shared in many ways among the decay products
"http://209.85.165.104/custom?q=cache:iuFso8Q0D4gJ:
www.lbl.gov/abc/wallchart/teachersguide/pdf/Chap03.pdf
+neutron+decay+product+energies&hl=en&ct=clnk&cd=16&gl=us&client
=pub-5123197681197851"

[40] Radiation Risk—ionization energy of a hydrogen atom is 13.6 eV
"http://hyperphysics.phy-astr.gsu.edu/hbase/nuclear/radrisk.html"

[41] Atom—Simple English Wikipedia, the free encyclopedia—atoms have a diameter of about 10^{-8} meters
"http://simple.wikipedia.org/wiki/Atom"

[42] Alpha Particle Tunneling
"http://hyperphysics.phy-astr.gsu.edu/hbase/nuclear/alptun.html"

[43] Quantum tunnelling—Wikipedia, the free encyclopedia
"http://en.wikipedia.org/wiki/Quantum_tunneling"

[44] Alpha Particle Tunneling—with less than 1/2 the kinetic energy to penetrate coulomb barrier.
"http://hyperphysics.phy-astr.gsu.edu/hbase/nuclear/alptun.html"

[45] Nuclear Binding Energy—alpha particle (helium nucleus) 28.3 Mev
"http://hyperphysics.phy-astr.gsu.edu/hbase/nucene/nucbin.html"

[46] Which is larger, the proton or the neutron—both about 1 femtometer
"http://www.physlink.com/education/AskExperts/ae570.cfm"

[47] Sun not hot enough to fuse two protons—tunneling required
"http://adsabs.harvard.edu/abs/2007APS.APRE15008B"

[48] Nuclear force—Wikipedia, the free encyclopedia—At much smaller separations between nucleons the force is very powerfully repulsive, which keeps the nucleons at a certain average separation.
"http://en.wikipedia.org/wiki/Nuclear_force"

[49] Atomic orbital—Wikipedia, the free encyclopedia—electron orbitals
"http://en.wikipedia.org/wiki/Atomic_orbital"

[50] The Shape of Molecules
"http://www.physchem.co.za/Bonding/Shape.htm"

[51] Pioneer anomaly—Wikipedia, the free encyclopedia
"http://en.wikipedia.org/wiki/Pioneer_anomaly"

[52] Sagnac effect Definition and Much More from Answers.com
"http://www.answers.com/topic/sagnac-effect"

[53] geocentric Definition and Much More from Answers.com—most educated Greeks from the 4th century BC on thought that the Earth was a sphere at the center of the universe.
"http://www.answers.com/topic/geocentric"

[54] The double-slit experiment (September 2002)—Physics World—PhysicsWeb "http://physicsweb.org/articles/world/15/9/1"

[55] Schrödinger's Atom—what is wrong with electron orbitals? "http://www.colorado.edu/physics/2000/quantumzone/schroedinger.html"

[56] Background Atoms and Light Energy—electrons radiate away excess energy "http://imagine.gsfc.nasa.gov/docs/teachers/lessons/xray_spectra/background-atoms.html"

[57] Science@Berkeley Lab Setting Free the Electrons—free electrons can radiate in burst as short as a 100 femtosecond. A femtosecond is a millionth of a billionth of a second.
"http://www.lbl.gov/Science-Articles/Archive/sabl/2007/Jan/free-electrons.html"

[58] "Photoelectric Effect." *Encyclopædia Britannica*. 2003. Encyclopædia Britannica Online. 24 Oct, 2003 http://www.search.eb.com/eb/article?eu=61313.

[59] Source: Britannica To cite this page: MLA style: "Radioactivity." *Encyclopædia Britannica*. 2003. Encyclopædia Britannica Online. 09 Nov, 2003 http://www.search.eb.com/eb/article?eu=119287.

[60] Laser Cooling—atoms are collected and then transferred into a purely optical trap formed by a focused CO_2 laser beam. The intensity of this laser beam is then turned down slowly, allowing the hottest atoms to escape—a process known as evaporative cooling.
"http://horology.jpl.nasa.gov/quantum/lasercooling.html"

[61] MLA style: "Atom." *Encyclopædia Britannica*. 2003. Encyclopædia Britannica Online. 10 Nov, 2003 http://www.search.eb.com/eb/article?eu=119284.

[62] Gravity as Curved Space Einstein's Theory of General Relativity—Space and space-time are not rigid arenas.
"http://theory.uwinnipeg.ca/mod_tech/node60.html"

[63] Michelson-Morley Experiment—from Eric Weisstein's World of Physics
"http://scienceworld.wolfram.com/physics/Michelson-MorleyExperiment.html"

[64] Scientific American Ask the Experts Physics—What exactly is the Higgs boson.
"http://www.sciam.com/askexpert_question.cfm?articleID= 00043456-7089-1C71-9EB7809EC588F2D7"

[65] The progress in the development of high-precision caesium beam atomic clocks gives the opportunity for detecting the hypothetical absolute velocity of the Earth in the terrestrial experiment, using the one-way light pulses.
"http://adsabs.harvard.edu/abs/2006EL....74.202K"

[66] Since the referential frame does not use any external reference, it can be taken as a universal frame from which absolute speeds can be defined.
"http://adsabs.harvard.edu/abs/2001EL....53.310S"

67] The Shell Model accounts for many features of the nuclear energy levels.
"http://www.lbl.gov/abc/wallchart/chapters/06/1.html"

[68] X-ray photons are associated with the **inner**-shell **electrons** close to the atomic nuclei, …
"http://www.britannica.com/nobelprize/article-59186"

[69] gamma-ray-induced fission of heavy isotopes …
"http://adsabs.harvard.edu/abs/1989Natur.337.718M"

[70] Neutron moderator—Wikipedia, the free encyclopedia
"http://en.wikipedia.org/wiki/Neutron_moderation"

[71] Madame Wu's experiment involved aligning the "http://everything2.com/index.pl?node=spin"s of a group of cobalt 60 nuclei and examining the direction in which the beta particles were emitted.
"http://everything2.com/index.pl?node=parity%20violation"

[72] Science News—reveal a slight positive charge at the neutron's center and a slight negative charge at its surface.
http://www.phschool.com/science/science_news/articles/not_so_neutral_neutron.html

73] BNEUTRON STRUCTURE and CLASSICAL RADIUS-B—*The neutron clearly shows a positive core (zone +) and a negative shell (zone—).*
"http://www.terra.es/personal/gsardin/news13.htm"

978-0-595-44092-
0-595-44092-4

Made in the USA
Columbia, SC
24 December 2017